Machine Learning

Machine Learning

Hands-On for Developers and Technical Professionals

Second Edition

Jason Bell

WILEY

Machine Learning: Hands-On for Developers and Technical Professionals, Second Edition

Copyright © 2020 by John Wiley & Sons, Inc., Indianapolis, Indiana

Published simultaneously in Canada

ISBN: 978-1-119-64214-5
ISBN: 978-1-119-64225-1 (ebk)
ISBN: 978-1-119-64219-0 (ebk)

Manufactured in the United States of America

For general information on our other products and services please contact our Customer Care Department within the United States at (877) 762-2974, outside the United States at (317) 572-3993 or fax (317) 572-4002.

Wiley publishes in a variety of print and electronic formats and by print-on-demand. Some material included with standard print versions of this book may not be included in e-books or in print-on-demand. If this book refers to media such as a CD or DVD that is not included in the version you purchased, you may download this material at http://booksupport.wiley.com. For more information about Wiley products, visit www.wiley.com.

Library of Congress Control Number: 2019956691

V10017436_020720

To all the developers who just wanted to get the code working without reading all the math stuff first.

About the Author

Jason Bell has worked in software development for more than 30 years. Currently he focuses on large-volume data solutions and helping retail and finance customers gain insight from data with machine learning. He is also an active committee member for several international technology conferences.

About the Technical Editor

Jacob Andresen works as a senior software developer based in Copenhagen, Denmark. He has been working as a software developer and consultant in information retrieval systems and web applications since 2002.

Acknowledgments

"Never again!" I think those were my final words after completing the first edition of this book. Five years later, and here we are again. When the call comes, you immediately think, "Well, it can't be hard, can it?"

To the Team

Jim Minatel, Devon Lewis, Janet Wehner, Pete Gaughan, and the rest of the team at Wiley, thank you for giving your blessing to this second edition and putting your faith in me to revise an awful lot of content. Apologies for the spelling mistakes and those *colour/color* occurrences. Many thanks to Jacob Andresen for giving a technical overview on the content of the book. His enthusiasm for the project was wonderful.

Most Excellent Friends and Collaborators

Dearest friends and acquaintances, thank you: Jennifer Michael, Marie Bentall, Tim Brundle, Stephen Houston, Garrett Murphy, Clare Conway, Tom Spinks, Matt Johnston, Alan Edwards, Colin Mitchell, Simon Hewitt, Mary McKenna, Alan Thorburn, Colin McHale, Dan Lyons, Victoria McCallum, Andrew Bolster, Eoin McFadden, Catherine Muldoon, Amanda Paver, Ben Lorica, Alastair Croll, Mark Madsen, Ellen Friedman, Ted Dunning, Sophia DeMartini, Bruce Durling, Francine Bennett, Michelle Varron, Elise Huard, Antony Woods, John Stephenson, McCraigMcCraig of the Clan McCraig, everyone on the Clojurians Slack Channel, the Strata Data community, Carla Gaggini, Kiki Schirr, Wendy

Devolder, Brian O'Neill, Anthony O'Connor, Tom Gray, Deepa Mann-Kler, Alan Hook, Michelle Douglas, Pete Harwood, Jen Samuel, and Colin Masters. There are loads I've forgotten, I know. I'm sorry.

And Finally

To my wife, Wendy, and my daughter, Clarissa, for absolutely everything and encouraging me to do these projects to the best of my nerdy ability. I couldn't have done it without you both.

To the rest of my family, Maggie, Fern, Andrew, Kerry, Ian and Margaret, William and Sylvia, thank you for all the support and kind words. William, if I need any more help, I'll call you.

The Bios That Never Made It. . .

"He has the boots and jacket that were the envy of many men."

"A dab hand at late-night YouTube videos of 80s pop stars."

"Jason Bell learned to play bass guitar on Saturday afternoons while pretending to work in a music shop."

Thanks to everyone who reads this book. I hope it's helpful in your journey. It's an honor and privilege that you chose to read it. Now I believe it's time for a cup of tea.

Contents

Introduction

Well, times have changed since writing the first edition of this book. Between 2014 and now there is more emphasis on data and what it can do for us but also how that power can be used against us. Hardware has gotten better, processing has gotten much faster, and the ability to classify, predict, and decide based on our data is extraordinary. At the same time, we've become much more aware of the risks of how data is used, the biases that can happen, and that a lot of black-box models don't always get things right.

Still, it's an exciting time to be involved. We still create more data than we can sensibly process. New ideas involving machine learning are being presented daily. The appetite for learning has grown rapidly, too.

Data mining and machine learning have been around a number of years already. When you look closely, the machine learning algorithms that are being applied aren't any different from what they were years ago; what is new is how they are applied at scale. When you look at the number of organizations that are creating the data, it's really, in my opinion, a minority. Google, Facebook, Twitter, Netflix, and a small handful of others are the ones getting the majority of mentions in the headlines with a mixture of algorithmic learning and tools that enable them to scale. So, the real question you should ask is, "How does all this apply to the rest of us?"

Data with large scale, near-instant processing, has come to the fore. The emphasis has moved from batch systems like Hadoop to more streaming-based systems like Kafka. I admit there will be times in this book when I look at the Big Data side of machine learning—it's a subject I can't ignore—but it's only a small factor in the overall picture of how to get insight from the available data. It is important to remember that I am talking about tools, and the key is figuring out which tools are right for the job you are trying to complete.

Aims of This Book

This book is about machine learning and not about Big Data. It's about the various techniques used to gain insight from your data. By the end of the book, you will have seen how various methods of machine learning work, and you will also have had some practical explanations on how the code is put together, leaving you with a good idea of how you could apply the right machine learning techniques to your own problems.

There's no right or wrong way to use this book. You can start at the beginning and work your way through, or you can just dip in and out of the parts you need to know at the time you need to know them.

"Hands-On" Means Hands-On

Many books on the subject of machine learning that I've read in the past have been very heavy on theory. That's not a bad thing. If you're looking for in-depth theory with really complex-looking equations, I applaud your rigor. Me? I'm more hands-on with my approach to learning and to projects. My philosophy is quite simple.

- Start with a question in mind.
- Find the theory I need to learn.
- Find lots of examples I can learn from.
- Put them to work in my own projects.

As a software developer, I like to see lots of examples. As a teacher, I like to get as much hands-on development time as possible but also get the message across to students as simply as possible. There's something about fingers on keys, coding away on your IDE, and getting things to work that's rather appealing, and it's something that I want to convey in the book.

Everyone has his or her own learning styles. I believe this book covers the most common methods, so everybody will benefit.

"What About the Math?"

Like arguing that your favorite football team is better than another or trying to figure out whether Jimmy Page is a better guitarist than Jeff Beck (I prefer Beck), there are some things that will be debated forever and a day. One such debate is how much math you need to know before you can start doing machine learning.

Doing machine learning and learning the theory of machine learning are two very different subjects. To learn the theory, a good grounding in math is required. This book discusses a hands-on approach to machine learning. With the number of machine learning tools available for developers now, the emphasis is not so much on how these tools work but on how you can make these tools work for you. The hard work has been done, and those who did it deserve credit and applause.

"But You Need a PhD!"

No, you don't!

The long-running debate rages on about the level of knowledge you need before you can start doing analysis on data or claim that you are a data scientist. I believe that if you'd like to take a few years completing a degree and then pursuing the likes of a master's degree and then a PhD, you should feel free to go that route. I'm a little more pragmatic about things and like to get reading and start doing.

Academia is great; and with the large number of online courses, papers, websites, and books on the subject of math, statistics, and data mining, there's enough to keep the most eager of minds occupied. I dip in and out of these resources a lot, and it's definitely a good way to keep up-to-date and investigate what's emerging.

For me, though, there's nothing like getting my hands dirty, grabbing some data, trying out some methods, and looking at the results. If you need to brush up on linear regression theory, then let me reassure you now, there's plenty out there to read, and I'll also cover that in this book.

Lastly, can one person ever be a data scientist? I think it's more likely for a team of people to bring the various skills needed for machine learning into an organization. I talk about this more in Chapter 2.

So, while others in the office are arguing whether to bring some PhD brains in on a project, you can be coding up a decision tree to see whether it's viable.

Over the last few years the job title data scientist has been joined by other titles like data engineer and machine learning engineer. All are valid and all focus on aspects of the data science pipeline. They all have their place.

What Will You Have Learned by the End?

Assuming that you're reading the book from start to finish, you'll learn the common uses for machine learning, different methods of machine learning, and how to apply real-time and batch processing.

There's also nothing wrong with referencing a specific section that you want to learn. The chapters and examples were created in such a way that there's no dependency to learn one chapter over another.

The aim is to cover the common machine learning concepts in a practical manner. Using the existing free tools and libraries that are available to you, there's little stopping you from starting to gain insight from the existing data that you have.

Balancing Theory and Hands-on Learning

There are many books on machine learning and data mining available, and finding the balance of theory and practical examples is hard. When planning this book, I stressed the importance of practical and easy-to-use examples, providing step-by-step instructions, so you can see how things are put together.

I'm not saying that the theory is light, because it's not. Understanding what you want to learn or, more importantly, how you want to learn will determine how you read this book.

You can think of the book split into three distinct sections. The first section covers the question, "What is machine learning?" and concentrates on planning for projects, data acquisition, and cleaning. For those wanting some refresher on the math and stats side of things, I've included a new chapter; it also covers linear regression and standard deviation.

The next section takes a closer look at some of the building-block algorithms used in machine learning projects. Clustering, decision trees, support vector machine, association rules learning, and neural networks provide both a background to how they work and code examples for you to work with. It's important to get the hands-on nature early on.

Lastly, I focus on the real-world tools used in enterprise; these are tools like Spark, Kafka, and R. Knowing how these frameworks and tools are put together will give you a grounding to know what to use when.

Source Code for This Book

All the code that is explained in the chapters of the book has been saved on a GitHub repository for you to download and try. For this edition, I've also included the Maven dependency file so you can easily build the project you are working on.

The address for the repository is `https://github.com/jasebell/mlbook2nd-edition`. You can also find it on the Wiley website at `www.wiley.com/go/machinelearning2e`.

The examples are in either Java, Clojure, or R. If you want to extend your knowledge into other languages, then a search around the GitHub site might lead you to some interesting examples.

Code has been separated by chapter; there's a folder in the repository for each of the chapters, and each has its own build file. The data is also within the repository in the data directory and has been split by each chapter.

Using Git

Git is a version control system that is widely used in business and the open source software community. If you are working in teams, it becomes useful because you can create branches of the codebase to work on then merge the changes afterward.

The uses for Git in this book are limited, but you need it for "cloning" the repository of examples if you want to use them.

To clone the examples for this book, use the following commands:

```
$mkdir mlbookexamples
$cd mlbookexamples
$git clone https://github.com/jasebell/mlbook2ndedition.git
```

You see the progress of the cloning, and when it's finished, you'll be able to change directories to the newly downloaded folder and look at the code samples.

What Is Machine Learning?

Let's start at the beginning, looking at what machine learning actually is, its history, and where it is used in industry. This chapter also describes some of the software used throughout the book so you can get everything installed and be ready to get working on the practical things.

History of Machine Learning

So, what is the definition of machine learning? Over the last six decades, several pioneers of the industry have worked to steer us in the right direction.

Alan Turing

In his 1950 paper, "Computing Machinery and Intelligence," Alan Turing asked, "Can machines think?" (For the full paper, see the link.)

```
www.csee.umbc.edu/courses/471/papers/turing.pdf
```

The paper describes the "Imitation Game," which involves three participants—a human acting as a judge, another human, and a computer that is attempting to convince the judge that it is human. The judge would type into a terminal program to "talk" to the other two participants. Both the human and the computer would respond, and the judge would decide which response came from the computer. If the judge couldn't consistently tell the difference between the human and computer responses, then the computer won the game.

The test continues today in the form of the Loebner Prize, an annual competition in artificial intelligence. The aim is simple enough: convince the judges that they are chatting to a human instead of a computer chat bot program.

Arthur Samuel

In 1959, Arthur Samuel defined machine learning as a field of study that "gives computers the ability to learn without being explicitly programmed." Samuel is credited with creating one of the first self-learning computer programs with his work at IBM. He focused on games as a way of getting the computer to learn things.

The game of choice for Samuel was checkers because it is a simple game but requires strategy from which the program could learn. With the use of alpha-beta evaluation pruning (eliminating nodes that do not need evaluating) and minimax strategies (minimizing the loss for the worst case), the program would discount moves and thus improve costly memory performance of the program.

Samuel is widely known for his work in artificial intelligence, but he was also noted for being one of the first programmers to use hash tables, and he certainly made a big impact at IBM.

Tom M. Mitchell

Tom M. Mitchell is the chair of machine learning at Carnegie Mellon University. As author of the book *Machine Learning* (McGraw-Hill, 1997), his definition of machine learning is often quoted.

> A computer program is said to learn from experience E with respect to some class of tasks T and performance measure P, if its performance at tasks in T, as measured by P, improves with the experience E.

The important thing here is that you now have a set of objects to define machine learning.

- Task (T), either one or more
- Experience (E)
- Performance (P)

So, with a computer running a set of tasks, the experience should be leading to performance increases.

Summary Definition

Machine learning is a branch of artificial intelligence. Using computing, we design systems that can learn from data in a manner of being trained. The systems might learn and improve with experience and, with time, refine a model that can be used to predict outcomes of questions based on the previous learning.

Algorithm Types for Machine Learning

There are a number of different algorithms that you can employ in machine learning. The required output is what decides which to use. As you work through the chapters, you'll see the different algorithm types being put to work. Machine learning algorithms characteristically fall into one of two learning types: supervised or unsupervised learning.

Supervised Learning

Supervised learning refers to working with a set of labeled training data. For every example in the training data you have an input object and an output object. An example would be classifying Twitter data. (Twitter data is used a lot in the later chapters of the book.) Assume you have the following data from Twitter; these would be your input data objects:

```
Really loving the new St Vincent album!
#fashion I'm selling my Louboutins! Who's interested? #louboutins
I've got my Kafka cluster working on a load of data. #data
```

For your supervised learning classifier to know the outcome result of each tweet, you have to manually enter the answers; for clarity, I've added the resulting output object at the start of each line.

```
music     Really loving the new St Vincent album!
clothing    #fashion I'm selling my Louboutins! Who's interested?
#louboutins
bigdata    I've got my Kafka cluster working on a load of data. #data
```

Obviously, for the classifier to make any sense of the data when run properly, you have to work manually on a lot more input data. What you have, though, is a training set that can be used for the later classification of data.

There are issues with supervised learning that must be taken into account. The *bias-variance dilemma* is one of them: how the machine learning model performs accurately using different training sets. High-bias models contain restricted learning sets, whereas high-variance models learn with complexity against noisy training data. There's a trade-off between the two models. The key is where to settle with the trade-off and when to apply which type of model.

Unsupervised Learning

On the opposite end of this spectrum is *unsupervised learning,* where you let the algorithm find a hidden pattern in a load of data. With unsupervised learning there is no right or wrong answer; it's just a case of running the machine learning algorithm and seeing what patterns and outcomes occur.

Unsupervised learning might be more a case of data mining than of actual learning. If you're looking at clustering data, then there's a good chance you're going to spend a lot of time with unsupervised learning in comparison to something like artificial neural networks, which are trained prior to being used.

The Human Touch

Outcomes will change, data will change, and requirements will change. Machine learning cannot be seen as a write-it-once solution to problems. Also, it requires human hands and intuition to write these algorithms. Remember that Arthur Samuel's checkers program basically improved on what the human had already taught it. The computer needed a human to get it started, and then it built on that basic knowledge. It's important that you remember that.

Throughout this book I talk about the importance of knowing what question you are trying to answer. The question is the cornerstone of any data project, and it starts with having open discussions and planning. (Read more about this in Chapter 2, "Planning for Machine Learning.")

It's only in rare circumstances that you can throw data at a machine learning routine and have it start to provide insight immediately.

Uses for Machine Learning

So, what can you do with machine learning? Quite a lot, really. This section breaks things down and describes how machine learning is being used at the moment.

Software

Machine learning is widely used in software to enable an improved experience with the user. With some packages, the software is learning about the user's behavior after its first use. After the software has been in use for a period of time, it begins to predict what the user wants to do.

Spam Detection

For all the junk mail that gets caught, there's a good chance a Bayesian classification filter is doing the work to catch it. From the early days of SpamAssassin to Google's work in Google Mail, there's been some form of learning to figure out whether a message is good or bad.

Spam detection is one of the classic uses of machine learning, and over time the algorithms have gotten better and better. Think about the e-mail program that you use. When it sees a message it thinks is junk, it asks you to confirm whether it is junk or isn't. If you decide that the message is spam, the system learns from that message and from the experience. Future messages will, ideally, be treated correctly from then on.

Voice Recognition

Apple's Siri service that is on many iOS devices is another example of software machine learning. You ask Siri a question, and it works out what you want to do. The result might be sending a tweet or a text message, or it could be setting a calendar appointment. If Siri can't work out what you're asking of it, it performs a Google search on the phrase you said.

Siri is an impressive service that uses a device and cloud-based statistical model to analyze your phrase and the order of the words in it to come up with a resulting action for the device to perform.

There's been a huge adoption of voice-activated assistants in the home like Amazon's Alexa and the Google Home device that take in voice commands and use machine learning to decide what the user is trying to do and come back with a response that is helpful.

Stock Trading

There are lots of platforms that aim to help users make better stock trades. These platforms have to do a large amount of analysis and computation to make recommendations. From a machine learning perspective, decisions are being made for you on whether to buy or sell a stock at the current price. It takes into account the historical opening and closing prices and the buy and sell volumes of that stock.

With four pieces of information (the low and high prices plus the daily opening and closing prices) a machine learning algorithm can learn trends for the stock. Apply this with all stocks in your portfolio, and you have a system to aid you in the decision whether to buy or sell.

Bitcoins are a good example of algorithmic trading at work; the virtual coins are bought and sold based on the price the market is willing to pay and the price at which existing coin owners are willing to sell.

The media is interested in the high-speed variety of algorithmic trading. The ability to perform many thousands of trades each second based on algorithmic prediction is a very compelling story. A huge amount of money is poured into these systems and how close they can get the machinery to the main stock trading exchanges. Milliseconds of network latency can cost the trading house millions in trades if they aren't placed in time.

About 70 percent of trades are performed by machine and not by humans on the trading floor. This is all very well when things are going fine, but when a problem occurs, it can be minutes before the fault is noticed, by which time many trades have happened. The flash crash in May 2010, when the Dow Jones industrial average dove 600 points, is a good example of when this problem occurred.

Robotics

Using machine learning, robots can acquire skills or learn to adapt to the environment in which they are working. Robots can acquire skills such as object placement, grasping objects, and locomotion skills through either automated learning or learning via human intervention.

With the increasing number of sensors within robotics, other algorithms could be employed outside of the robot for further analysis.

We can't talk about robotics without mentioning the self-driving car. Huge strides have been made since the first edition of this book. Tesla has the autopilot feature enabling the car to self-drive while the driver is still close by with hands near the wheel. It's still in the early days, and there is the obvious discussion about job displacement and the resulting new job creation.

Medicine and Healthcare

The race is on for machine learning to be used in healthcare analytics. A number of startups are looking at the advantages of using machine learning with Big Data to provide healthcare professionals with better-informed data to enable them to make better decisions.

IBM's famed Watson supercomputer, once used to win the television quiz program *Jeopardy* against two human contestants, is being used to help doctors. Using Watson as a service on the cloud, doctors can access learning on millions of pages of medical research and hundreds of thousands of pieces of information on medical evidence.

With the number of consumers using smartphones and the related devices for collating a range of health information—such as weight, heart rate, pulse, pedometers, blood pressure, and even blood glucose levels—it's now possible to track and trace user health regularly and see patterns in dates and times. Machine learning systems can recommend healthier alternatives to the user via the device.

Image processing has gotten more powerful, and it's becoming easier to diagnose via X-ray and MRI scans to detect various cancers and other disease pointers.

Although it's easy enough to analyze data, protecting the privacy of user health data is another story. Obviously, some users are more concerned about how their data is used, especially in the case of it being sold to third-party companies. The increased volume of analytics in healthcare and medicine is new, but the privacy debate will be the deciding factor about how the algorithms will ultimately be used.

Advertising

For as long as products have been manufactured and services have been offered, companies have been trying to influence people to buy their products. Since 1995, the Internet has given marketers the chance to advertise directly to our screens without needing television or large print campaigns. Remember the thought of cookies being on our computers with the potential to track us? The race to disable cookies from browsers and control who saw our habits was big news at the time.

Log file analysis is another tactic that advertisers use to see the things that interest us. They are able to cluster results and segment user groups according to who may be interested in specific types of products. Couple that with mobile location awareness and you have highly targeted advertisements sent directly to you.

There was a time when this type of advertising was considered a huge invasion of privacy, but we've gradually gotten use to the idea, and some people are even happy to "check in" at a location and announce their arrival. If you're thinking your friends are the only ones watching, think again. In fact, plenty of companies are learning from your activity. With some learning and analysis, advertisers can do a good job of figuring out where you'll be on a given day and attempt to push offers your way.

Retail and E-commerce

Machine learning is heavily used in retail, both in e-commerce and in bricks-and-mortar retail. At a high level, the obvious use case is the loyalty card. Retailers that issue loyalty cards often struggle to make sense of the data that's coming back to them. Because I worked with one company that analyzes this data, I know the pain that supermarkets go through to get insight.

UK supermarket giant Tesco is the leader when it comes to customer loyalty programs. The Tesco Clubcard is used heavily by customers and gives Tesco a great view of customer purchasing decisions. Data is collected from the point of sale (POS) and fed back to a data warehouse. In the early days of the Clubcard, the data couldn't be mined fast enough; there was just too much. As processing methods improved over the years, Tesco and marketing company Dunn Humby have developed a good strategy for understanding customer behavior and shopping habits and encouraging customers to try products similar to their usual choices.

An American equivalent is Target, which runs a similar sort of program that tracks every customer engagement with the brand, including mailings, website visits, and even in-store visits. From the data warehouse, Target can fine-tune how to get the right communication method to the right customers in order for them to react to the brand. Target learned that not every customer wants an e-mail or an SMS message; some still prefer receiving mail via the postal service.

The uses for machine learning in retail are obvious: Mining baskets and segmenting users are key processes for communicating the right message to the customer. On the other hand, it can be too accurate and cause headaches. Target's "baby club" story, which was widely cited in the press as a huge privacy danger in Big Data, showed us that machine learning can easily determine that we're creatures of habit, and when those habits change, they will get noticed.

TARGET'S PRIVACY ISSUE

Target's statistician, Andrew Pole, analyzed basket data to see whether he could determine when a customer was pregnant. A select number of products started to show up in the analysis, and Target developed a pregnancy prediction score. Coupons were sent to customers who were predicted to be pregnant according to the newly mined score. That was all very well until the father of a teenage girl contacted his local store to complain about the baby coupons that were being sent to his daughter. It turned out that Target predicted the girl's pregnancy before she had told her father that she was pregnant.

For all the positive uses of machine learning, there are some urban myths, too. For example, you might have heard the "beer and diapers" story associated with Walmart and other large retailers. The idea is that the sales of beer and diapers both increase on Fridays, suggesting that mothers were going out and dads would stock up on beer for themselves and diapers for the little ones they were looking after. It turned out to be a myth, but this still doesn't stop marketing companies from wheeling out the story (and believing it's true) to organizations who want to learn from their data.

Another myth is that the heavy-metal band Iron Maiden would mine Bit-Torrent data to figure out which countries were illegally downloading their songs and then fly to those locations to play concerts. That story got the marketers and media very excited about Big Data and machine learning, but sadly it's untrue. That's not to say that these things can't happen someday; they just haven't happened yet.

Gaming Analytics

We've already established that checkers is a good candidate for machine learning. Do you remember those old chess computer games with the real plastic pieces? The human player made a move, and then the computer made a move. Well, that's a case of machine learning planning algorithms in action. Fast-forward a few decades (the chess computer still feels like yesterday to me) to today when the console market is pumping out analytics data every time you play your favorite game.

Microsoft has spent time studying the data from *Halo 3* to see how players perform on certain levels and also to figure out when players are using cheats. Fixes have been created based on the analysis of data coming back from the consoles. Other games producers like Blizzard (*Overwatch*), Epic Games (*Fortnite*), and Respawn Entertainment (*Apex Legends*) use large matrix calculations to ensure that players are suitably matched before a game can start.

Microsoft also worked on Drivatar, which is incorporated into the driving game *Forza Motorsport*. When you first play the game, it knows nothing about your driving style. Over a period of practice laps the system learns your style, consistency, exit speeds on corners, and positioning on the track. The sampling happens over three laps, which is enough time to see how your profile behaves. As time progresses, the system continues to learn from your driving patterns. After you've let the game learn your driving style, the game opens up new levels and lets you compete with other drivers and even your friends.

Even within story-based games like *The Last of Us* by Naughty Dog, characters within gameplay scenes are aware of their surroundings and other characters within the gameplay. For example, if a bottle is thrown and smashes, enemies, friends, and infected alike would be alerted and their next moves decided by in-play artificial intelligence.

If you have children, you might have seen the likes of *Nintendogs* (or cats), a game in which a person is tasked with looking after an on-screen pet. (Think Tamagotchi, but on a larger scale.) Algorithms can work out when the pet needs to play, how to react to the owner, and how hungry the pet is.

It's still the early days of game companies putting machine learning into infrastructure to make the games better. With more and more games appearing on small devices, such as those with the iOS and Android platforms, the real learning is in how to make players come back and play more and more. Analysis can be performed about the "stickiness" of the game—do players return to play again, or do they drop off over a period of time in favor of something else? Ultimately there's a trade-off between the level of machine learning and gaming performance, especially in smaller devices. Higher levels of machine learning require more memory within the device. Sometimes you have to factor in the limit of what you can learn from within the game.

The Internet of Things

Connected devices that can collate all manner of data are sprouting up all over the place. Device-to-device communication is hardly new, but it hadn't really hit the public minds until fairly recently. With the low cost of manufacture and distribution, now devices are being used in the home just as much as they are in industry.

Uses include home automation, shopping, and smart meters for measuring energy consumption. These things are in their infancy, and there's still a lot of concern about the security aspects of these devices. In the same way mobile device location is a concern, companies can pinpoint devices by their unique IDs and eventually associate them to a user.

On the plus side, the data is so rich that there's plenty of opportunity to put machine learning in the heart of the data and learn from the devices' output. This may be as simple as monitoring a house to sense ambient temperature—for example, is it too hot or too cold?

Languages for Machine Learning

This book uses the Java and Clojure programming languages for the working examples. The reasons are simple: Java is a widely used language, especially in the enterprise, and the libraries are well supported. Clojure gives better data handling abilities thanks to its functional nature: data goes into a function, and the result is output as data. Java isn't the only language to be used for machine learning—far from it. If you're working for an existing organization, you may be restricted to the languages used within it.

With most languages, there is a lot of crossover in functionality. With the languages that access the Java Virtual Machine (JVM) there's a good chance that you'll be accessing Java-based libraries. There's no such thing as one language being "better" than another. It's a case of picking the right tool for the job. The following sections describe some of the other languages that you can use for machine learning.

Python

The Python language has increased in usage because it's easy to learn and easy to read. It also has some good machine learning libraries, such as scikit-learn, PyML, and pybrain. Jython was developed as a Python interpreter for the JVM, which may be worth investigating.

If you are looking at the Tensorflow libraries, then Python is an obvious choice, and while there are Java extensions available, I'd recommend using Python in the first instance.

R

R is an open source statistical programming language. The syntax is not the easiest to learn, but I do encourage you to take a look at it. It also has a large number of machine learning packages and visualization tools. The RJava project allows Java programmers to access R functions from Java code. For a basic introduction to R, take a look at Chapter 14, "Machine Learning with R."

Matlab

The Matlab language is used widely within academia for technical computing and algorithm creation. Like R, it also has a facility for plotting visualizations and graphs.

Scala

A new breed of languages is emerging that takes advantage of Java's runtime environment, which potentially increases performance, based on the threading architecture of the platform. Scala (which is an acronym for *Sca*lable *La*nguage) is one of these, and it is being widely used by a number of startups.

There are machine learning libraries, such as ScalaNLP, but Scala can access Java JAR files, and it can also implement the likes of Classifier4J and Mahout. It's also core to the Apache Spark project, which is covered in Chapter 13, "Apache Spark"

Ruby

Many people know about the Ruby language by association with the Ruby on Rails web development framework, but it's also used as a stand-alone language. The best way to integrate machine learning frameworks is to look at JRuby, which is a JVM-based alternative that enables you to access the Java machine learning libraries.

Software Used in This Book

The hands-on elements in the book use a number of programs and packages to get the algorithms and machine learning working.

To keep things easy, I strongly advise that you create a directory on your system to install all these packages. I'm going to call mine `mlbook`.

```
$mkdir ~/mlbook
$cd ~/mlbook
```

Checking the Java Version

As the programs used in the book rely on Java, you need to quickly check the version of Java that you're using. The programs require Java 1.8, or newer. To check your version, open a terminal window and run the following:

```
$ java -version
java version "1.7.0_40"
Java(TM) SE Runtime Environment (build 1.7.0_40-b43)
Java HotSpot(TM) 64-Bit Server VM (build 24.0-b56, mixed mode)
```

If you are running a version older than 1.6, then you need to upgrade your Java version. You can download the current version from

www.oracle.com/technetwork/java/javase/downloads/index.html

Weka Toolkit

Weka (Waikato Environment for Knowledge Acquisition) is a machine learning and data mining toolkit written in Java by the University of Waikato in New Zealand. It provides a suite of tools for learning and visualization via the supplied workbench program or the command line. Weka also enables you to retrieve data from existing data sources that have a JDBC driver. With Weka you can do the following:

- Preprocessing data
- Clustering
- Classification
- Regression
- Association rules

The Weka toolkit is widely used and now supports the Big Data aspects by interfacing with Hadoop for clustered data mining.

You can download Weka from the University of Waikato website at

www.cs.waikato.ac.nz/ml/weka/downloading.html

There are versions of Weka available for Linux, macOS, and Windows. To install Weka on Linux, you just need to unzip the supplied file to a directory. On macOS and Windows, an installer program is supplied that will unzip all the required files for you.

DeepLearning4J

For the more involved neural networks, I'll be using the DeepLearning4J library. As it is written in Java, it scales well, but even better is that it can use Spark for some of its preprocessing. This means it can scale with Big Data where other languages might struggle.

In this book I'm using the required libraries within the Maven dependency file (pom.xml), but if you want to read more about what DeepLearning4J can do, then please visit https://deeplearning4j.org.

Kafka

In the first edition of the book I made the decision to use SpringXD as the data ingestion engine. Since then, Kafka has proven itself to be a market leader when it comes to streaming data. There are two community editions that you can download; there's the Apache Kafka distribution and the community edition from Confluent, which is the commercial arm of Kafka.

For the examples in this book and especially in Chapter 12, "Machine Learning Streaming with Kafka," where I use Kafka for self-training machine learning applications, I'll be using the Apache Kafka distribution.

Spark and Hadoop

Customers using Hadoop are still out there, but they are beginning to be treated in the same way that legacy databases are. Spark made inroads within the Big Data community, and it's becoming the de facto processing framework for data at scale.

In this book, I'll be using version 2.4.4 against the Hadoop 2.7 binaries. For more information on Spark, please visit https://spark.apache.org.

Text Editors and IDEs

Some discussions seem to spark furious debate in certain circles—for example, favorite actor/actress, best football team, and best integrated development environment (IDE).

I now use the IntelliJ IDEA Java development platform for my Java-based development. For Clojure development, I use Emacs with a host of packages installed. For a look at the Emacs packages, see the Home Light Sabre Kit on my GitHub account.

```
https://github.com/jasebell/home-lightsaber-kit
```

It's a fork from Bruce Durling's original project.

Data Repositories

One question that comes up again and again in my classes is "Where can I get data?" There are a few answers to this question, but the best answer depends on what you are trying to learn.

Data comes in all shapes and sizes, which is something discussed in the next chapter. I strongly suggest you take some time to hunt around the Internet for different datasets and look through them. You'll get a feel for how these things are put together. Sometimes you'll find comma-separated variable (CSV) data, or you might find JSON or XML data.

Remember, some of the best learning comes from playing with the data. Having a question in mind that you are trying to answer with the data is a good start (and something you will see me refer to a number of times in this book), but learning comes from experimentation and improvement on results. So, I'm all for playing around with the data first and seeing what works. I hail from a pragmatic background when it comes to development and learning. Although the majority of publications about machine learning have come from people with academic backgrounds—and I fully endorse and support them—we shouldn't discourage learning by doing.

The following sections describe some places where you can get plenty of data with which to play.

UC Irvine Machine Learning Repository

This machine learning repository consists of more than 270 datasets. Included in these sets are notes on the variable name, instances, and tasks the data would be associated with. You can find this repository at `http://archive.ics.uci.edu/ml/datasets`.

Kaggle

The competitions that Kaggle runs have gained a lot of interest over the last couple of years. The 101 section on the site offers some datasets with which to experiment. You can find them at `www.kaggle.com/competitions`.

Summary

This chapter looked at what machine learning is, how it can be applied to different areas of business, and what tools you need to follow along with the remainder of the book.

The next chapter introduces you to planning for machine learning. It covers data science teams, cleaning, and different methods of processing data.

Planning for Machine Learning

This chapter looks at planning your machine learning projects, storage types, processing options, and data input. The chapter also covers data quality and methods to validate and clean data before you do any analysis.

The Machine Learning Cycle

A machine learning project is basically a cycle of actions that need to be performed (see Figure 2.1).

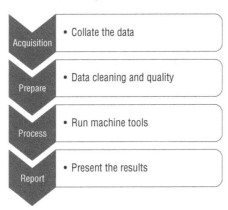

Figure 2.1: The machine learning process

You can acquire data from many sources; it might be data that's held by your organization or open data from the Internet. There might be one dataset, or there could be 10 or more.

You must come to accept that data will need to be cleaned and checked for quality before any processing can take place. These processes occur during the prepare phase.

The processing phase is where the work gets done. The machine learning routines that you have created perform this phase.

Finally, the results are presented. Reporting can happen in a variety of ways, such as reinvesting the data into a data store or reporting the results as a spreadsheet or report.

It All Starts with a Question

There seems to be a misconception that machine learning, like Big Data, is a case of throwing enough data at the problem that the answers magically appear. As much as I'd like to say this happens all the time, it doesn't. Machine learning projects start with a question or a hunch that needs investigating. I've encountered this quite a few times in speaking to people about their companies' data ambitions and what they are looking to achieve with the likes of machine learning and Hadoop.

Using a whiteboard, sticky notes, or even a sheet of paper, start asking questions like the following:

- Is there a correlation between our sales and the weather?
- Do sales on Saturday and Sunday generate the majority of revenue to the business compared to the other five days of the week?
- Can we plan what fashions to stock in the next three months by looking at Twitter data for popular hashtags?
- Can we tell when our customers become pregnant?

All these examples are reasonable questions, and they also provide the basis for proper discussion. Stakeholders will usually come up with the questions, and then the data project team (which might be one person—you!) can spin into action.

Without knowing the question, it's difficult to know where to start. Anyone who thinks the answers just pop out of thin air needs a polite, but firm, explanation of what has to happen for the answers to be discovered.

I Don't Have Data!

This sounds like a silly statement when you have a book on machine learning in your hands, but sometimes people just don't have the data.

In an ideal world, we expect companies to have well-groomed customer relationship management (CRM) systems and neat repositories of data that could be retrieved on a whim and copied nicely into a Hadoop filesystem, so countless MapReduce jobs could run (read more about MapReduce in Chapter 13, "Apache Spark and MLLib").

Data comes from a variety of sources. Plenty of open data initiatives are available, so you have a good chance of being able to find some data to work with.

Starting Local

Perhaps you could make a difference in your local community; see what data they have open with which you can experiment. New York City has a whole portal of open data with more than 1,100 datasets for citizens to download and learn from. Hackathons and competitions encourage people to get involved and give back to the community. The results of the hackathons make a difference because insights about how the local community is run are fed back to the event organizers. If you can't find the dataset you want, then you are also encouraged to request it.

Transfer Learning

With the amount of machine learning now being executed out in the field, it may be worth looking into existing models and altering certain parameters to fit in with your prediction data, especially if you don't have much in the way of training data. This is called *transfer learning*. It's perfect for models that require large scale datasets for training, such as images, video, and large text corpus. I'll highlight some transfer learning examples in later chapters.

Competitions

If you fancy a real challenge, then think about entering competitions. One of the most famous was the Netflix Prize, which was a competition to improve the recommendation algorithm for the Netflix film service.

Teams that were competing downloaded sample sets of user data and worked on an algorithm to improve the predictions of movies that customers would like. The winning team was the one that improved the results by 10 percent. In 2009, the $1 million prize was awarded to "BellKor's Pragmatic Chaos." This triggered a new wave of competitions, letting the data out into the open so collaborative teams could improve things.

In 2010, Anthony Goldbloom founded `Kaggle.com`, which is a platform for predictive modeling and analytics competitions. Each competition posted has sample datasets and a brief of the desired outcome. Either teams or individuals can enter, and the most effective algorithms, similar to the Netflix Prize, decide the winner.

Is competition effective? It seems to be. Kaggle has more than 100,000 data scientists registered from across the world. Organizations such as Facebook, NASA, GE, Wikipedia, and AllState have used the service to improve their products and even head-hunt top talent.

One Solution Fits All?

Machine learning is built up from a varying set of tools, languages, and techniques. It's fair to say that there is no one solution that fits most projects. As you will find in this chapter and throughout the book, I'll refer to various tools to get certain aspects of the job done. For example, there might be data in a relational database that needs extracting to a file before you can process it.

Over the last few years, I've seen managers and developers with faces of complete joy and happiness when a data project is assigned. It's new, it's hip, and, dare I say it, it's funky to be working on data projects. Then after the scale of the project comes into focus, I've seen the color drain from their faces. Usually this happens after the managers and developers see how many different elements are required to get things working for the project to succeed. And, like any major project, the specification from the stakeholders will change things along the way.

Defining the Process

Making anything comes down to process, whether that's baking a cake, brewing a cup of coffee, or planning a machine learning project. Processes can be refined as time goes on, but if you've never developed one before, then you can use the following process as a template.

Planning

During the late 1980s, I wrote many assignments and papers on the upcoming trend of the paperless office and how computers would one day transform the way day-to-day operations would be performed. Even without the Internet, it was easy to see that computers were changing how things were being done.

Skip ahead to the present day and you'll see that my desk is littered with paper, notebooks, sticky notes, and other scraps of information. The paperless office didn't quite make the changes I was expecting, and you need no more evidence than the state of my desk. I would show you a photograph, but it might prove embarrassing.

What I have found is that all projects start on paper. For me, it doesn't work to jump in and code; I find that method haphazard and error prone. I need to plan first. I use A5 Moleskin notebooks for notes and use A4 and A3 artist drawing pads for large diagrams. They're on my desk, in my bag, and in my jacket pocket.

Whiteboards are good, too. Whiteboards hold lots of ideas and diagrams, but I find they can get out of control and messy after a while. There was once an office wall in Santa Clara that I covered in sticky notes. (I did take them down once I was finished. The team thought I was mad.)

Planning might take into account where the data is coming from, if it needs to be cleaned, what learning methods to use, and what the output is going to look like. The main point is that these things can be changed at any time—the earlier in the process they change, the better. So, it's worth taking the time to sit around a table with stakeholders and the team and figure out what you are trying to achieve.

Developing

This process might involve algorithm development or code development. The more iterations you perform on the code, the better it will be. Agile development processes work best; in agile development, you work only on what needs to be done without trying to future-proof the software as you go along. It's worth using some form of code repository site like GitHub or Bitbucket to keep all your work private; it also means you can roll back to earlier versions if you're not happy with the way things are going.

Testing

In this case, testing means testing with data. You might use a random sample of the data or the full set. The important thing is to remind yourself that you're testing the process, so it's okay for things to not go as planned. If you push things straight to production, then you won't really know what's going to happen. With testing you can get an idea of the pain points. You might find data-loading issues, data-processing issues, or answers that just don't make sense. When you test, you have time to change things.

Reporting

Sit down with the stakeholders and discuss the test results. Do the results make sense? The developers and mathematicians might want to amend algorithms or the code. Stakeholders might have a new question to ask (this happens a lot), or perhaps you want to introduce some new data to get another angle on the answers. Regardless of the situation, make sure the original people from the planning phase are back around the table again.

Refining

When everyone is happy with the way the process is going, it's time to refine code and, if possible, the algorithms. With huge volumes of data, if you squeeze every ounce of performance you can from your code, the quicker the overall processing time will be. Think of a bobsled run; a slower start converts to a much slower finish.

Production

When all is tested, reviewed, and refined by the team, moving to production shouldn't be a big job. Be sure to give consideration to when this project will be run—is it an hourly/daily/weekly/monthly job? Will the data change wildly between the project going into production and the next run?

Make sure the team reviews the first few production runs to ensure the results are as expected and then look at the project as a whole to see whether it's meeting the criteria of the stakeholders. Things might need to be refined. As you probably already know, software is rarely finished.

Avoiding Bias

Let's not forget that machine learning is hard; getting models that avoid bias is hard. It's important to get the teams talking to each other about how to avoid introducing any form of bias into the final solution.

Dataset choice is important. Make sure that it's evenly weighted. Whether it's gender, age, location, or another parameter, if the quantity of that source data type is too heavy, then your model is going to bias toward it.

Try using different model types and evaluate the training and the test predictions. Unsupervised models can introduce bias with tighter correlations when clustering from the training data, and supervised models can creep in bias when human intervention is brought in to control either the model or the training data. Care must be taken.

Building a Data Team

A *data scientist* is someone who can bring the facets of data processing, analytics, statistics, programming, and visualization to a project. With so many skill sets in action, even for the smallest of projects, it's a lot to ask for one person to have all the necessary skills. In fact, I'd go as far as to say that such a person might not exist—or is at least extremely rare. A data science team might touch on some, or all, of the following areas of expertise.

Mathematics and Statistics

Someone on the team needs to have a good head for mathematics—someone who isn't going to flinch when the words *linear regression* are mentioned in the interview. I'm not saying there's a minimum level of statistics you should know before embarking on any project, but knowledge of descriptive statistics (the mean, the mode, and the median), distributions, and outliers will give you a good grounding to start.

The debate will rage on about the level of mathematics needed in any machine learning project, but my opinion is that every project comes with its own set of complications. If new information needs to be learned, then there are plenty of sources out there from which you can learn.

If you have access to talented mathematicians, then your data team is a blessed group indeed.

Programming

Good programming talent is hard to come by, but I'm assuming that if you have this book in your hand, then there's a good chance you're a programmer already. Taking algorithms and being able to transfer that to workable code can take time and planning. It's also worth knowing some of the Big Data tools, such as the Spark framework and Kafka. (Read Chapters 12 and 13 for a comprehensive walk-through on both technologies.)

Graphic Design

Visualizing data is important; it tells the story of your findings to the stakeholders or end users. Although much emphasis has been placed on the Web for presentation with technologies such as D3 and Processing, don't forget the likes of BIRT, Jasper Reports, and Crystal Reports.

This book doesn't touch on visualization, but Appendix D, "Further Reading," includes some titles that will point you in the right direction.

Domain Knowledge

If, for example, you are working with medical data, then it would be beneficial to have someone who knows the medical field well. The same goes for retail; there's not much point trawling through rows of transactions if no one knows how to interpret how customers behave. Domain experts are the vital heroes in guiding the team through a project. There are some decisions that the domain expert will instinctively know.

Think of a border crossing with passport control. There might be many permutations of rules that are given depending on nationality, immigration rules, and so on. A domain expert would have this knowledge in place and make your life as the developer much easier and would help to get a solution up and running more quickly.

There's a notion that we don't need domain experts. I'm of the mind that we do, even if you only sit down and have coffee with someone who knows the domain. Always take a notebook and keep notes.

Data Processing

After you have a team in place and a rough idea of how all of this is going to get put together, it's time to turn your attention to what is going to do all the work for you. You must give thought to the frequency of the data process jobs that will take place. If it will occur only once in a while, then it might be false economy investing in hardware over the long term. It makes more sense to start with what you have in hand and then add as you go along and as you notice growth in processing times and frequency.

Using Your Computer

Yes, you can use your own machine, either a desktop or a laptop. I do my development on an Apple MacBook Pro. I run the likes of Kafka, Spark, and Hadoop on this machine as it's pretty fast, and I'm not using terabytes of data. There's nothing to stop you from using your own machine; it's available, and it saves financial outlay to get more machines. Obviously, there can be limitations. Processing a heavy job might mean you have to turn your attention to less processor-intensive things, but never rule out the option of using your own machine.

Operating systems like Linux and macOS tend to be preferred over Windows, especially for Big Data–based operations. The best choice comes down to what you know best and what suits the project best in order to get the job done efficiently. I don't believe there's only one right way to do things.

A Cluster of Machines

Eventually you'll come across a scenario that requires you to use a cluster of machines to do the work. Frameworks like Hadoop are designed for use over clusters of machines, which make it possible for the distribution of work to be done in parallel. Ideally the machines should be on the same network to reduce network traffic latency.

At this point in time, it's also worthwhile to add a good system administrator to the data science team. Any performance that can be improved over the cluster will bring a marked performance improvement to the whole project.

Cloud-Based Services

If the thought of maintaining and paying for your own hardware does not appeal, then consider using some form of cloud-based service. Vendors such as Amazon, Rackspace, and others provide scalable servers where you can increase, or decrease, the number of machines and amount of power that you require.

The advantage of these services is that they are "turn on/turn off" technology, enabling you to use only what you need.

Keep a close eye on the cost of cloud-based services, as they can sometimes prove more expensive than just using a standard hosting option over longer time periods. Some companies provide dedicated Big Data services if you require the likes of Spark to do your processing. With cloud-based services, it's always important to turn the instance off; otherwise, you'll be charged for the usage while the instance is active.

Data Storage

There are some decisions to make on how the data is going to be stored. This might be on a physical disc or deployed on a cloud-based solution.

Physical Discs

The most common form of storage is the one that you will more than likely have in your computer to start off with. The hard disc is adequate for testing and small jobs. You will notice a difference in performance between physical discs and solid-state drives (SSDs); the latter provides much faster performance. External drives are cheap, too, and provide a good storage solution for when data volumes increase.

Cloud-Based Storage

Plenty of cloud-based storage facilities are available to store your data as required. If you are looking at cloud-based processing, then you'll more than likely be purchasing some form of cloud-based storage to go with it. For example, if you use Amazon's Elastic Map Reduce (EMR) system, then you would be using it alongside the S3 storage solution; other storage solutions exist for the Microsoft Azure platform and Google Cloud Compute.

Like cloud processing, storage based on the cloud will cost you on a monthly or annual basis. You also have to think about the bandwidth implications of moving large volumes of data from your office location to the cloud system, which is another cost to keep in mind.

Data Privacy

Data is power and with it comes an awful lot of responsibility. The privacy issue will always rage on in the hearts and minds of the users and the general public. Everyone has an opinion on the matter, and often people err on the side of caution.

In the last five years there has been a huge emphasis on how user data is used. In Europe, the General Data Protection Regulations control how business can use personal data within their organizations.

Ultimately, with great power comes great responsibility, and it will be up to you how that data is protected and processed.

Cultural Norms

Cultural expectations are difficult to measure. As the World Wide Web has progressed since the mid-1990s, there has been a privacy battle about everything from how cookies were stored on your computer to how a multitude of companies are tracking locations, social interactions, ratings, and purchasing decisions through your mobile devices.

If you're collecting data via a website or mobile application, then there's an expectation that you will be giving something in return for user information. When you collect that information, it's only right to tell the user what you intend to do with the data.

Supermarket loyalty card schemes are a simple data-collecting exercise. For every basket that goes through the checkout, there's the potential that the customer has a loyalty card. In associating that customer with that basket of products you can start to apply machine learning. Over time you will be able to see the shopping habits of that customer—her average spend, the day of the week she shops—and the customer expects some form of discount promotion for telling you all this information.

So, how do you keep cultural norms onside? By giving customers a clear opt-in or opt-out strategy.

Generational Expectations

During sessions of my iPhone development class, I open up with a discussion about personal data. I can watch the room divide instantly, and I can easily see the deciding factor: age.

Some people are more than happy to share with their friends, and the rest of the world, their location, what they are doing, and with whom. These people post pictures of their activities and tag them so they could be easily searched, rated, and commented on. They use Facebook, Instagram, YouTube, Twitter, and other apps as a normal, everyday part of their lives.

The other group of people, who were older, are not comfortable with the concept of handing over personal information. Some of them think that no one in their right minds would be interested in such information. Most can't see the point.

Although the generation gap might be closing and there is a steady relaxation of what people are willing to put on the Internet, developers have a responsibility to the suppliers of the information. You have to consider whether the results you generate will cause a concern to them or enhance their lives.

The Anonymity of User Data

You can learn from data, but users get touchy when their names are attached to it. Creating hashes of important data is a starting point, but it's certainly not the end game. Consider my name as an MD5 hash. Using the Linux `md5sum` command, I can find it out easily, as shown here:

```
$ printf '%s' "Jason Bell" | md5sum
a7b19ed2ca59f8e94121b54f9f26333c  -
```

Now, I have a hash value, which is a good start, but it's still not really protecting my identity. You now know it and what it would possibly relate to if it were used as a user key in a machine learning process. It wouldn't take much time for a decent programmer with a list of first and last names to generate all the `md5` values for all the combinations.

Using a salt value is a better solution. A *salt value* is random data that's used with the piece of data to make it more secure and harder to crack.

Let's assume the salt value is the number of nanoseconds from January 1, 1970. You take that and the string you're looking to hash.

```
$ printf '%s' "Jason Bell $(date +%sN)" | md5sum
40e46b48a873c30c80469dbbefaa5e16  -
```

There are different ways of handling the input string. You might want to remove spaces, but the concept remains the same. The security of these hashes has to be maintained by you so when the time comes to interpret the answers, you'll know which customers are doing the actions you are seeking. Hashes aren't just restricted to usernames or customer names; they can be applied to any data. Anything that you consider private information (known as *personally identifiable information* [PII])—something that you don't want any third party to see—must be hashed.

Don't Cross the "Creepy Line"

Be careful not to make the customer freak out by crossing the line in the sand that I call the "creepy line." It's the point where the horrified customer would shriek, "How did they know that?" For an example of a company and what they know about you, visit the settings pages of your Google account and take a look at your web search history or your location history:

```
https://www.google.com/settings/dashboard
```

One near-legendary example in data science, Big Data, and machine learning circles is the story of Target and pregnant mothers, which was widely cited on the Internet because of Charles Duhigg's book *The Power of Habit* (Random

House, 2011). What readers of the Internet forgot to realize was that Target had been using the same practice for years; the concept was originally run in 2002 as an exercise to see if there was a correlation between two things.

Good mathematics and item matching isolated a number of items that mothers-to-be started to buy. Target has enough data to predict what trimester of the pregnancy the mother is in. With an opt-in to the baby club, this might have all passed without problem. But when an angry father rolls up to the store to inquire why his teenage daughter is receiving baby promotions and coupons, well, that's a different matter.

What does this example highlight? Well, apart from freaking out the customer, it causes undue pressure on the in-store staff. Everyone in the organization needs to be aware of the work that's going on. Also, the data team needs to be acutely aware of the social effect of their learning.

The UK supermarket chain Tesco started the Clubcard loyalty scheme in 1995; it holds more data than some governments on customer purchasing behavior, social classes, and income bracket. The store's data processing power is controlled by a marketing company, Dunn Humby, which runs the Clubcard and analyzes the data. What is the upside for the customer? Four times a year Clubcard members receive coupons for money off and incentives to buy items they normally purchase. The offers resemble the customers' typical shopping patterns, but other items are thrown in so it doesn't look like they've been stalked.

Mining the baskets is hardly a new idea (you'll be reading about other techniques in later chapters), but when the supermarket becomes large and the volumes of data are huge, the insight that can be gained becomes an enormous commercial advantage. The cost of this advantage is appearing to know the intimate shopping details of the customer even when they've not overtly given permission for you to send offers.

Data Quality and Cleaning

In an ideal world, you'd receive data and put it straight into the system for processing. Then your favorite actor or actress would hand you your favorite drink and pat you on the back for a job well done.

In the real world, data is messy, usually unclean, and error prone. The following sections offer some basic checks you should do, and I've included some sample data so you can see clearly what to look for.

The example data is a simple address book with a first name, last name, e-mail address, and age.

Presence Checks

First things first, check that data has been entered at all. Within web-based businesses, registration usually involves at least an e-mail address, first name, and last name. It's amazing how many times users will try to avoid putting in their names.

The presence check is simple enough. If the field length is empty or null and that piece of data is important in the analysis, then you can't use records from which the data is missing.

	FIRSTNAME	LASTNAME	E-MAIL	AGE
Correct	Jason	Bell	me@domain.com	42
Incorrect		Bell		42

The first name and e-mail are missing from the example, so the record should really be fixed or rejected. In theory, the data could be used if knowing the customer was not important.

Type Checks

With relational databases you have schemas created, so there's already an expectation of what type of data is going where. If incorrect data is written to a field of a different data type, then the database engine will throw an error and complain at you.

In text data, such as CSV files, that's not the case, so it's worth looking at each field and ensuring that what you're expecting to see is valid.

```
#firstname, lastname, email, age
Jason,Bell,me@domain.com,42
42,Bell,me@domain.com,Jason
```

From the example, you can see that the first row of data is correct, but the second is wrong because the `firstname` field has a number in it and not a string type. There are a couple of things you could do here. The first option is to ignore the record, as it doesn't fit the data-quality check. The other option is to see if any other records have the same e-mail address and check the name against those records.

Length Checks

Field lengths must be checked, too; once again, relational databases exercise a certain amount of control, but textual data can be error-prone if people don't go with the general rules of the schema.

FIELD	LENGTH	GOOD	BAD
Firstname	10	Jason	Mr Jason Bell
Email	20	me@domain.com	jason.bell@thing.domain.com

Range Checks

Range or reasonableness checks are used with numeric or date ranges. Age ranges are the main talking point here. Until there are advances in scientific medicine to prolong life, you can make a fairly good assumption that the upper lifespan of someone is about 120. You can even play it safe and extend the upper range to 150; anyone who is older than that is lying or just trying to put a false value in to trip up the system.

FIELD	LOWER RANGE	UPPER RANGE
Age	0	120
Month	1	12

Format Checks

When you know that certain data must follow a given format, then it's always good to check it. Regular expression knowledge is a big advantage here if you know it. E-mail addresses can be used and abused in web forms and database tables, so it's always a good idea to validate what you can at the source.

There's much discussion in the developer world about what a correct e-mail regular expression actually is. The official standard for the e-mail address specification is RFC 5322. Correctly matching the e-mail address as a regular expression is a huge pattern. What you're looking for is something that will catch the majority of e-mail addresses.

```
[a-z0-9!#$%&'*+/=?^_`{|}~-]+(?:\.[a-z0-9!#$%&'*+/=?^_`{|}~-]+)*@ (?:[a-
z0-9](?:[a-z0-9-]*[a-z0-9])?\.)+[a-z0-9](?:[a-z0-9-]*[a-z0-9])?
```

The main thing to do is create a run of test cases with all the eventualities of an e-mail address you think you will come across. Don't just test it once; keep retesting it over time.

Postcodes and ZIP codes are another source of formatting woe—especially UK postcodes. Regular expressions also help in this case, but sometimes an odd one slips through the testing. At the end of the day, this sort of thing is better left to specialized software or expert services.

The Britney Dilemma

Users being users will input all sorts of things, and it's really up to us to make sure that our software catches what it can. Although search strings aren't specific to machine learning, it is, however, an interesting case of how different names can really mess up the results.

For instance, take the variations of the search term *Britney Spears* in a well-known search engine. In an ideal and slightly utopian vision, everyone would type her name perfectly into a text field box.

```
britney spears
```

Life rarely goes as planned, and users type what they think is right, such as the following:

```
brittany spears
brittney spears
britany spears
britny spears
briteny spears
britteny spears
briney spears
brittny spears
brintey spears
britanny spears
britiny spears
britnet spears
britiney spears
britaney spears
britnay spears
brithney spears
brtiney spears
birtney spears
brintney spears
briteney spears
bitney spears
brinty spears
brittaney spears
brittnay spears
britey spears
brittiny spears
```

If you were to put that through a Hadoop cluster looking for unique singer search terms, you'd be in a bit of a mess, as each of these would register a new result count.

What you want is something to weigh each term and see what it resembles. The simplest approach is to use a classifier to weigh each search term as it comes in. You know the correct term, so it's a case of running the incoming terms against the correct one and seeing what the confidence scoring is.

```
package mlbook.ch02.examples;

import java.util.ArrayList;
import java.util.List;
```

```
import net.sf.classifier4J.ClassifierException;
import net.sf.classifier4J.vector.HashMapTermVectorStorage;
import net.sf.classifier4J.vector.TermVectorStorage;
import net.sf.classifier4J.vector.VectorClassifier;

public class BritneyDilemma {

    public BritneyDilemma() {
        List<String> terms = new ArrayList<String>();
        terms.add("brittany spears");
        terms.add("brittney spears");
        terms.add("britany spears");
        terms.add("britny spears");
        terms.add("briteny spears");
        terms.add("britteny spears");
        terms.add("briney spears");
        terms.add("brittny spears");
        terms.add("brintey spears");
        terms.add("britanny spears");
        terms.add("britiny spears");
        terms.add("britnet spears");
        terms.add("britiney spears");
        terms.add("christina aguilera");

        TermVectorStorage storage = new HashMapTermVectorStorage();
        VectorClassifier vc = new VectorClassifier(storage);
        String correctString = "britney spears";

        for (String term : terms) {
          try {
            vc.teachMatch("sterm", correctString);
            double result = vc.classify("sterm", term);
            System.out.println(term + " = " + result);
          } catch (ClassifierException e) {
            e.printStackTrace();
          }
        }
    }

    public static void main(String[] args) {
        BritneyDilemma bd = new BritneyDilemma();
    }
}
```

This code sample uses the Classifer4J library to run a basic vector space search on the incoming spellings of Britney; it then ranks them against the correct string. When this code is run, you get the following output:

```
brittany spears = 0.7071067811865475
brittney spears = 0.7071067811865475
```

```
britany spears = 0.7071067811865475
britny spears = 0.7071067811865475
briteny spears = 0.7071067811865475
britteny spears = 0.7071067811865475
briney spears = 0.7071067811865475
brittny spears = 0.7071067811865475
brintey spears = 0.7071067811865475
britanny spears = 0.7071067811865475
britiny spears = 0.7071067811865475
britnet spears = 0.7071067811865475
britiney spears = 0.7071067811865475
britaney spears = 0.7071067811865475
britnay spears = 0.7071067811865475
brithney spears = 0.7071067811865475
brtiney spears = 0.7071067811865475
birtney spears = 0.7071067811865475
brintney spears = 0.7071067811865475
briteney spears = 0.7071067811865475
bitney spears = 0.7071067811865475
brinty spears = 0.7071067811865475
brittaney spears = 0.7071067811865475
brittnay spears = 0.7071067811865475
britey spears = 0.7071067811865475
brittiny spears = 0.7071067811865475
christina aguilera = 0.0
```

The confidence is always a number between 0 and 0.9999. Just to prove that, putting the correct spelling in the list and running the program again would generate a positive score.

```
britney spears = 0.9999999999999998
```

Obviously, there's some preparation required, as you need to know the correct spellings of the search terms before you can run the classifier. This example just proves the point.

What's in a Country Name?

Data cleaning needs to be done in a variety of circumstances, but the most common reason is too many options were given in the first place.

A few years ago, I was looking at a database for a hotel. Its data was gathered via a web-based inquiry form, but instead of offering a selection of countries from a drop-down list of countries, there was just an open text field. (Always remember that freedom of input, where it can be avoided, should be avoided.)

Let's consider this for a moment. If you take a country like Ireland, then you might have the following entries for country name:

- Ireland
- Republic of Ireland
- Eire
- EIR
- Rep. of Ireland

All these are essentially the same place; the only exception would be Northern Ireland, which is still part of the United Kingdom.

What you have is a huge job to clean up the country field of a database. To fix this, you would have to find all the distinct names in the country field and associate them with a two-letter country code. So, Ireland and all the other names that were associated with Ireland become IE. You would have to do this for all the countries. Where possible, it's better to have tight control of the input data, as this will make things a lot easier when it comes to processing.

In programming terms, you could make each of the distinct countries a key in a `HashMap` and add a method to get the value of the corresponding input name.

```java
package mlbook.ch02.examples;

import java.util.HashMap;
import java.util.Map;

public class CountryHashMap {

    private Map<String, String> countries = new HashMap<String, String>();

    public CountryHashMap() {
        countries.put("Ireland", "IE");
        countries.put("Eire", "IE");
        countries.put("Republic of Ireland", "IE");
        countries.put("Northern Ireland", "UK");
        countries.put("England", "UK");
        // you could add more or generate from a database.
    }

    public String getCountryCode(String country) {
        return countries.get(country);
    }

    public static void main(String[] args) {
        CountryHashMap chm = new CountryHashMap();
        System.out.println(chm.getCountryCode("Ireland"));
        System.out.println(chm.getCountryCode("Northern Ireland"));
    }
}
```

The preceding example is a basic piece of code that would automate the cleaning process in a short amount of time. However, you are strongly advised to look at the source of the problem and refactor the input. If no change is made, then the same cost to the business will occur, as you'll have to clean the data again.

Ideally, to avoid having to do this sort of cleaning, you would employ verification strategies at the input stage. So, for example, if you're using web forms, you should use JavaScript to validate the input before it's saved to the database. Other times you inherit data and occasionally have to employ such methods.

Dates and Times

For time series processing, you must ensure that you have a consistent set of dates to read. The format you choose is really up to you. International Standard ISO 8601 lays out the specification for date and time representations in a numerical format. The issue with the ISO 8601 standard is that it's not immune to the Y10K bug when timestamps will be incorrect after January 19, 2038. The Temps Atomique International (TAI) standard takes into account these issues.

Regardless of the language you are using, make yourself aware of how the date formatting and parsing routines work. For Java, take a look at the SimpleDateFormat API, which gives you a rundown on all the settings along with some useful examples. Use caution when running code on distributed systems and also with different time zones.

Table 2.1 shows some of the commonly used date/time formats.

Table 2.1: Commonly Used Date/Time Formats

DATE/TIME FORMAT	SIMPLEDATEFORMAT REPRESENTATION
2014-01-01	Yyyy-MM-dd
2014-01-01 11:59:00	Yyyy-MM-dd hh:mm:ss
1388577540	(Unix timestamps are like long variable types but with nano seconds added.)

I've seen many a database table with different date formats that have been saved as string types. Things have gotten better, but it's still something I keep in mind.

Final Thoughts on Data Cleaning

Data cleaning is a big deal, because it increases the chances of getting better results. For some Big Data projects, 80 percent of the project time is spent on data cleaning before the actual analysis starts. It's important to keep this step high up in the project plan and manage time accordingly.

Thinking About Input Data

With any machine learning project, you need to think about the incoming data, what format it's in, and how it will be accessed by the code that's being built.

Data comes in all sorts of forms, so it's a good idea to know what you're dealing with before you start crafting any code. The following sections describe some of the more common data formats.

Raw Text

Basic raw text files are used in many publications. If you look at the likes of the Guttenberg Project, you'll see that you can download works in a raw text file. The data is unstructured, so it rarely has a proper form with which you can work.

```
Lorem ipsum dolor sit amet, consectetur adipiscing elit. Suspendisse
eget metus quis erat tempor hendrerit. Vestibulum turpis ante, bibendum
vitae nisi non, euismod blandit dui. Maecenas tristique consectetur est
nec elementum. Maecenas porttitor, arcu sed gravida tempus, purus tellus
lacinia erat, dapibus euismod felis enim eget nisl. Nunc mollis volutpat
ligula. Etiam interdum porttitor nulla non lobortis.
```

Common formats for text files are Unicode, ASCII, or UTF-8. If there's any international encoding required, UTF-8 and Unicode are most common. Note that PDF documents, Rich Text Format files, and Word documents are not raw text files. Microsoft Office documents (such as Word files) are particularly troublesome because of "smart quotes" and other nontext extraneous characters that wreak havoc in Java programs.

Comma-Separated Variables

The CSV format is widely used across the data landscape. The comma character is used between each field of data. You might find that other delimiters are used, such as tabulation (TSV) and the pipe (|) symbol (PSV). Delimiters are not limited to one character either. If you look at something like the USDA Food Database, you'll see ~^~ used as a delimiter. The following CSV file is generated from a fake name generator site. (It's always good to use fake data when you're testing things.)

```
1,male,Mr.,Joe,L,Perry,50 Park Row,EDERN,,LL53 2SQ,GB,United
Kingdom,JoePerry@einrot.com,Annever,eiThahph9Ah,077 6473
7650,Fry,7/4/1991,Visa,4539148712302735,342,2/2018,YB 20 98
60 A,1Z 23F 389 61 4167 727 1,Blue,Nephrology nurse,Friendly
Advice,1999 Alfa Romeo 145,BadProtection.co.uk,O+,169.4,77.0,5'
10",177,a617f840-6e42-4146-b743-090ee59c2c9f,52.806493,-4.72918
```

```
2,male,Mr.,Daniel,J,Carpenter,51 Guildford Rd,EAST
DRAYTON,,DN22 3GT,GB,United Kingdom,DanielCarpenter@teleworm.
us,Reste1990,Eich1Kiegie,079 2890 2948,Harris,3/26/1990,MasterCard,
5353722386063326,717,7/2018,KL 50 03 59 C,1Z 895 362 50 0377 620
2,Blue,Corporate administrative assistant,Hit or Miss,2000 Jeep Grand
Cherokee,BiologyConvention.co.uk,AB+,175.3,79.7,5' 7",169,ac907a59-a091-
4ba2-9b0f-a1276b3b5ada,52.801024,-0.719021

3,male,Mr.,Harvey,A,Hawkins,37 Shore Street,STOKE TALMAGE,,OX9
4FY,GB,United Kingdom,HarveyHawkins@armyspy.com,Spicionly,UcheeGh9xoh,077
7965 0825,Rees,3/1/1974,MasterCard,5131613608666799,523,7/2017,SS 81 32
33 C,1Z Y11 884 19 7792 722 8,Black,Education planner,Monsource,1999 BMW
740,LightingShadows.co.uk,A-,224.8,102.2,6' 1",185,6cf865fb-81ae-42af-
9a9d-5b86d5da7ce9,51.573674,-1.179834

4,male,Mr.,Kyle,E,Patel,97 Cloch Rd,ST MARTIN,,TR12 6LT,GB,United
Kingdom,KylePatel@superrito.com,Wilvear,de2EeJew,079 2879 6351,Hancock,
6/7/1978,Visa,4916480323599950,960,4/2016,MH 93 02 76 D,1Z 590 692
15 4564 674 8,Blue,Interior decorator,Grade A Investment,2002 Proton
Juara,ConsumerMenu.co.uk,AB+,189.2,86.0,5' 10",179,e977c58e-ba61-406e-
a1d1-2904807be365,49.957435,-5.258628

5,male,Mr.,Dylan,A,Willis,66 Temple Way,WINWICK,,WA2 5HE,GB,United
Kingdom,DylanWillis@cuvox.de,Hishound,shael7Foo,077 1105 4178,Kelly,
8/16/1948,Visa,4485311140499796,423,11/2016,WG 24 10 62 D,1Z 538 4E0
39 8247 102 7,Black,Community health educator,Mr. Steak,2002 Nissan
X-Trail,FakeRomance.co.uk,A+,170.1,77.3,5' 9",175,335c2508-71be-43ad-
9760-4f5c186ec029,53.443749,-2.631634

6,female,Mrs.,Courtney,R,Jordan,42 Kendell Street,SHARLSTON,,WF4
1PZ,GB,United Kingdom,CourtneyJordan@fleckens.hu,Ponforsittle,
Hi2oteel1,070 3469 5710,Payne,2/23/1982,MasterCard,55708
15007804057,456,12/2019,CJ 87 95 98 D,1Z 853 489 84
8609 859 3,Blue,Mechanical inspector,Olson Electronics,2000
Chrysler LHS,LandscapeCovers.co.uk,B+,143.9,65.4,5'
3",161,27d229b0-6106-4700-8533-5edc2661a0bf,53.645118,-1.563952
```

People might refer to files as CSV files even though they are not comma separated. The best way to find out if something is really a CSV file is to open up the data and take a look.

JSON

JavaScript Object Notation (JSON) is a commonly used data format that utilizes key-value pairs to communicate data between machines and the Web. It was designed as an alternative to XML. Don't be fooled by the use of the word

JavaScript; you don't need JavaScript to use this data format. There are JSON parsers for various languages. The earlier CSV example used fake name data; here's the first entry of the CSV in JSON notation:

```
[
  {
    "Number":1,
    "Gender":"male",
    "Title":"Mr.",
    "GivenName":"Joe",
    "MiddleInitial":"L",
    "Surname":"Perry",
    "StreetAddress":"50 Park Row",
    "City":"EDERN",
    "State":"",
    "ZipCode":"LL53 2SQ",
    "Country":"GB",
    "CountryFull":"United Kingdom",
    "EmailAddress":"JoePerry@einrot.com",
    "Username":"Annever",
    "Password":"eiThahph9Ah",
    "TelephoneNumber":"077 6473 7650",
    "MothersMaiden":"Fry",
    "Birthday":"7/4/1991",
    "CCType":"Visa",
    "CCNumber":4539148712302735,
    "CVV2":342,
    "CCExpires":"2/2018",
    "NationalID":"YB 20 98 60 A",
    "UPS":"1Z 23F 389 61 4167 727 1",
    "Color":"Blue",
    "Occupation":"Nephrology nurse",
    "Company":"Friendly Advice",
    "Vehicle":"1999 Alfa Romeo 145",
    "Domain":"BadProtection.co.uk",
    "BloodType":"O+",
    "Pounds":169.4,
    "Kilograms":77.0,
    "FeetInches":"5' 10\"",
    "Centimeters":177,
    "GUID":"a617f840-6e42-4146-b743-090ee59c2c9f",
    "Latitude":52.806493,
    "Longitude":-4.72918
  }
]
```

Many application programming interfaces (APIs) use JSON to send response data back to the requesting program. Some parsers might take the JSON data and represent it as an object. Others might be able to create a hash map of the data for you to access.

YAML

Whereas JSON is a document markup format, YAML (meaning "YAML Ain't Markup Language") is most certainly a data format. It's not as widely used as JSON but from a distance looks similar.

```
date    : 2014-01-02
bill-to: &id001
    given  : Jason
    family : Bell
    address:
        lines: |
            458 Some Street Somewhere
            In Some Suburb
        city   : MyCity
        state  : CA
        postal : 55555
```

XML

The Extensible Markup Language (XML) followed on from the popular use of Standard Generalized Markup Language (SGML) for document markup. The idea was for XML to be easily read by humans and also by machines. On first inspection, XML is like Hypertext Markup Language (HTML); later versions of HTML use strict XML formatting types.

XML gets criticism for its complexity, especially when reading large structures. That's one reason it's popular for web-based APIs to use JSON data as its response. There are a large number of APIs delivering XML response data, so it's worthwhile to look at how it works:

```xml
<?xml version="1.0" encoding="UTF-8" ?>
    <Customer>
        <Number>1</Number>
        <Gender>male</Gender>
        <Title>Mr.</Title>
        <GivenName>Joe</GivenName>
        <MiddleInitial>L</MiddleInitial>
        <Surname>Perry</Surname>
        <StreetAddress>50 Park Row</StreetAddress>
        <City>EDERN</City>
        <State></State>
        <ZipCode>LL53 2SQ</ZipCode>
        <Country>GB</Country>
        <CountryFull>United Kingdom</CountryFull>
        <EmailAddress>JoePerry@einrot.com</EmailAddress>
        <Username>Annever</Username>
        <Password>eiThahph9Ah</Password>
```

```
        <TelephoneNumber>077 6473 7650</TelephoneNumber>
        <MothersMaiden>Fry</MothersMaiden>
        <Birthday>7/4/1991</Birthday>
        <CCType>Visa</CCType>
        <CCNumber>4539148712302735</CCNumber>
        <CVV2>342</CVV2>
        <CCExpires>2/2018</CCExpires>
        <NationalID>YB 20 98 60 A</NationalID>
        <UPS>1Z 23F 389 61 4167 727 1</UPS>
        <Color>Blue</Color>
        <Occupation>Nephrology nurse</Occupation>
        <Company>Friendly Advice</Company>
        <Vehicle>1999 Alfa Romeo 145</Vehicle>
        <Domain>BadProtection.co.uk</Domain>
        <BloodType>O+</BloodType>
        <Pounds>169.4</Pounds>
        <Kilograms>77</Kilograms>
        <FeetInches>5' 10&quot;</FeetInches>
        <Centimeters>177</Centimeters>
        <GUID>a617f840-6e42-4146-b743-090ee59c2c9f</GUID>
        <Latitude>52.806493</Latitude>
        <Longitude>-4.72918</Longitude>
    </Customer>
```

Most of the common languages have XML parsers available using either a document object model (DOM) parser or the Simple API for XML (SAX) parser. Both types come with advantages and disadvantages depending on the size and complexity of the XML document with which you are working.

Spreadsheets

Talk to any finance person in your organization, and you'll discover that their entire world revolves around spreadsheets. Programmers have a tendency to shun spreadsheets in favor of data formats that make their lives easier. You can't totally ignore them, though. Spreadsheets are the lifeblood of an organization, and they probably hold most of the organization's data.

There are lots of different spreadsheet programs, but the most commonly used applications are Microsoft Excel, Google Docs Spreadsheet, and LibreOffice.

Fortunately, there are programming APIs that you can use to extract the data from spreadsheets directly, which saves a lot of work in converting the spreadsheet to the likes of CSV files. It's worth studying the formulas in the spreadsheets, because there might be some algorithms lurking there that are worth their weight in gold.

If you want your finance person to be supportive of the project, tell that person that the results will be in a spreadsheet and you'll have a friend for a long time after.

The Java programming language has a few APIs to choose from that will enable you to read and write spreadsheets. The Apache POI project and JExcel API are the two most popular.

Databases

If you've been brought up with web programming, then you might have had some exposure to databases and database tables. Common ones are MySQL, Postgres, Microsoft SQL Server, and Oracle.

Recently, there's been an explosion of NoSQL (meaning Not Only SQL), such as MongoDB, CouchDB, Cassandra, Redis, and HBase, which all bring their own flavors to data storage. These document and key-value stores move away from the rigid table-like structures of traditional databases.

In addition, there are graph databases such as Apache Giraph and Neo4J and in-memory systems such as Spark, memcached, and Storm. Chapter 13 is an introduction to Spark.

In my opinion, all databases have their place and are worth investigating. There's nothing wrong with having relational, document, and graph databases running concurrently for the project. Each has its advantages to the project that you might not have considered. As with all these things, there might be a learning curve that you need to factor into your project time.

Images

The common data formats previously mentioned mainly deal with text or numbers in different shades, but you can't discount images. There are a number of things you can learn from images. Whether you're trying to use facial recognition or emotion tracking or you're trying to determine whether an image is a cat or dog (yes, it has been done), there are several APIs that will help.

The most popular formats are Portable Network Graphics (PNG) and JPEG images; these are regularly used on the Web. If processing power is freely available, then TIFF or BMP are much larger files, but they contain more image information.

Ultimately our job is to convert images to numbers so the algorithms can work with the vectors of number information. This will require reducing image size and then doing the conversion. More of these techniques are covered in Chapter 11, "Machine Learning from Image Information."

Thinking About Output Data

Now it's time to turn your attention to the output data. This is where the stakeholders might have a say in how things are going to be done, because ultimately it will be those people who deal with the results.

The primary question about the output of machine learning data is "Who is the intended audience?" Depending on the answer to that question, your output will vary. You might need a spreadsheet for the financial folks to see the results. If the audience is comprised of website users, then it makes sense to put the data back into a database table. The machine learning results could be merged with other data to define more learning. It really comes down to what was defined in the project.

There are a number of paid and free reporting tools available. Some are full-blown systems, such as Jasper Reports, BIRT, and Tableau. If you are reporting to a web-based audience, then the likes of D3 and Processing might be of help to you.

Don't Be Afraid to Experiment

It's safe to say that there is no "one solution fits all." There are many components, formats, tools, and considerations to ponder on any project. In effect, every machine learning project starts with a clean sheet and communication among all involved, from stakeholders all the way through to visualization. Tools and scripts can be reused, but every case is going to be different, so things need minor adjustments as you go along. Don't be afraid to play around with data as you acquire it; see whether there's anything you can glean from it.

It's also worth taking time to grab some open data and make your own scenarios and ask your own questions. It's like a musician practicing an instrument; it's worth putting in the hours so you are ready for the day when the big gig arrives.

The machine learning community is large, and there are plenty of blog posts, articles, videos, and books produced by the community. Forums are the perfect place to swap stories and experiences, too. As with most things, the more you put in, the more you will get out of it.

Over the years, I've found that people are more than willing to help contribute to a solution if you're stuck on a problem. If you haven't looked at the likes of `http://stackoverflow.com`, a collaborative question-and-answer platform for software developers, then have a search around. Chances are that someone will have encountered the same problem as you.

Summary

As with any project, planning is a key and essential part of machine learning and shouldn't be taken lightly. This chapter covered many aspects of planning, including processing, storage, privacy, and data cleaning. You were also introduced to some useful tools and commands that will help in the cleaning phases and some validation checks.

The planning phase is a constantly evolving process, and the more machine learning projects you and the team perform, the more you will learn from previous mistakes.

The key is to start small. Take a snapshot of the data and take a random sample with a size of 10 percent of the total. Get the team to inspect the data. Can you work with it? Do you anticipate any problems with the processing of this data?

Cleaning the data might take the most time of the project; the actual processing might consume only a fraction of the overall project time. If you can supply clean data, then your results will be refined.

Regardless of whether you are working on a 10-person team or on your own, be aware of your network of contacts; some might have domain knowledge that will be useful. Ask lots of questions, too. You'd be surprised how many folks are willing to answer questions in order to see you succeed.

The next few chapters examine some different machine learning techniques and put some sample code together, so you can start to apply them to your own projects.

Data Acquisition Techniques

"Computers aren't the thing. They're the thing that gets us to the thing."

—Joe MacMillan

This quote comes from the television program *Halt and Catch Fire*; perhaps we should reconsider that statement for our purposes: "Data isn't the thing. Data is the thing that gets us to the thing." The question to ask is where is the data coming from and does it need cleaning or transforming?

When it comes to machine learning and machine learning projects, you'll spend a large portion of your time on getting the data into the right shape so it can be processed. Welcome to the dark art that is extracting, transforming, and loading data.

Scraping Data

The sad fact of reality is that data is rarely neatly packaged the way we want. Sure, there are exceptions like WikiData and the Facebook Graph API, and there are application programming interfaces (APIs) that will give you nicely prepared data (more on that shortly). But you must be prepared to work with the messy world of scraping data.

Processing scraped data requires a few steps to get it from the usual messy state it's in to something usable.

1. Figure out where the data is coming from.
2. Figure out how you're going to get it.
3. Make it machine readable.
4. Make sure the values are workable.
5. Figure out where to store it.

Copy and Paste

There will be a day you'll have to extract data from a web page or a series of web pages. Truth be told, they tend to be a mess, but some are better than others. A first attempt would be to copy and paste the data from the page and then figure a way out to remove the HTML tags. There are, however, easier ways. Let's look at an example.

Suppose we've been tasked with extracting airport data. I'd like to see the busiest airports in the United Kingdom. I've found a page on Wikipedia, and I'd like to get the data (see Figure 3.1).

Figure 3.1: Wikipedia list of the busiest airports in United Kingdom

The link to visit is here:

```
https://en.wikipedia.org/wiki/List_of_busiest_airports_in_the_United_
Kingdom
```

There are several tables that have the information I'm looking for. For this example, I want to look at the 2017–2018 figures. If I were to copy/paste the 2017–2018 into a text file, the output is okay but needs cleaning (see Figure 3.2).

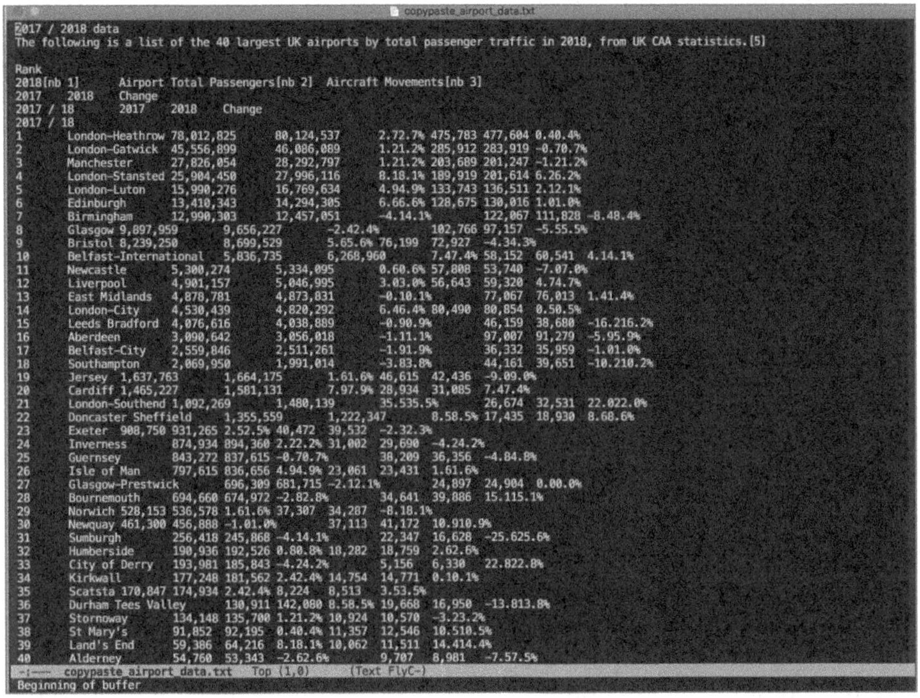

Figure 3.2: Text file of 2017–2018 data

The actual data doesn't start until line 9. Fortunately, the copy and paste that I've done has preserved the tab characters, but it does require some work. I can run a command-line operation and apply a regular expression on the data to convert the tabs to pipes so I have a visual reference for the columns.

I'm using Perl to do the search and replace. Then to inspect the results, I use the `head` command, which will display the first 20 lines of the output.

```
$ cp copypaste_airport_data.txt copypaste_airport_data_piped.txt
$ perl -i -p -e "s/\t/\|/g;" copypaste_airport_data_piped.txt
$ head -n 20 copypaste_airport_data_piped.txt
2017 / 2018 data
The following is a list of the 40 largest UK airports by total passenger
traffic in 2018, from UK CAA statistics.[5]

Rank
2018[nb 1]|Airport|Total Passengers[nb 2]|Aircraft Movements[nb 3]
2017|2018|Change
2017 / 18|2017|2018|Change
2017 / 18
```

```
1|London-Heathrow|78,012,825|80,124,537|2.72.7%|475,783|477,604|0.40.4%
2|London-Gatwick|45,556,899|46,086,089|1.21.2%|285,912|283,919|-0.70.7%
3|Manchester|27,826,054|28,292,797|1.21.2%|203,689|201,247|-1.21.2%
4|London-Stansted|25,904,450|27,996,116|8.18.1%|189,919|201,614|6.26.2%
5|London-Luton|15,990,276|16,769,634|4.94.9%|133,743|136,511|2.12.1%
6|Edinburgh|13,410,343|14,294,305|6.66.6%|128,675|130,016|1.01.0%
7|Birmingham|12,990,303|12,457,051|-4.14.1%|122,067|111,828|-8.48.4%
8|Glasgow|9,897,959|9,656,227|-2.42.4%|102,766|97,157|-5.55.5%
9|Bristol|8,239,250|8,699,529|5.65.6%|76,199|72,927|-4.34.3%
10|Belfast-International|5,836,735|6,268,960|7.47.4%|58,152|
60,541|4.14.1%
11|Newcastle|5,300,274|5,334,095|0.60.6%|57,808|53,740|-7.07.0%
12|Liverpool|4,901,157|5,046,995|3.03.0%|56,643|59,320|4.74.7%
```

Let's review that Perl script again.

```
$ perl -i -p -e "s/\t/\|/g;" copypaste_airport_data_piped.txt
```

The flags set up things for us. The -i flag sets the output of the script to the same as the filename that was read. It's worth working on a backup copy of the source data. If it all goes wrong, then you can copy the source file again and give it another go. An input loop is constructed around the script with -p, and the -e flag is to enter a single line of script, that being the regular expression.

The regular expression is a simple search and replace.

```
"s/<replace this>/<with this>/g;"
```

At the end of the expression is g;, which means applying it globally to the entire string.

Going back to the output, that Perl script seems to have worked! I'm excited now and a little bit closer to getting the data I need. However, on inspection, I start to see issues with the data. Looking at the first row, I see things like 2.72.7% and 0.40.4%, so there's a data issue. I could hand edit them to the correct values, but that's time intensive. Or I could craft another regular expression, but that could create errors that are then difficult to pick up. The more processes you add to parse or fix your data, the more chance you have to add errors to the resulting output. I'm now at the point where I want another approach.

Google Sheets

The spreadsheet program that Google supplies has a function that not many people talk about. So, I'll let you in on the secret.

Create a new sheet from the main Drive menu. Once you get the blank spreadsheet, type in the following formula command in the first cell (A1):

```
=importhtml("https://en.wikipedia.org/wiki/List_of_busiest_airports_in_
the_United_Kingdom","table",1)
```

The function is in three parts. The first is the URL that you want to load into the spreadsheet. Second, there's the entity type you want to extract; in this example, it's the table. The last part is the instance of the entity to extract. For the airports, it's the first table I want.

You'll see the first cell change to "Loading" while the spreadsheet fetches the page, and after a few seconds, the data will appear all nice and neat in the spreadsheet (see Figure 3.3).

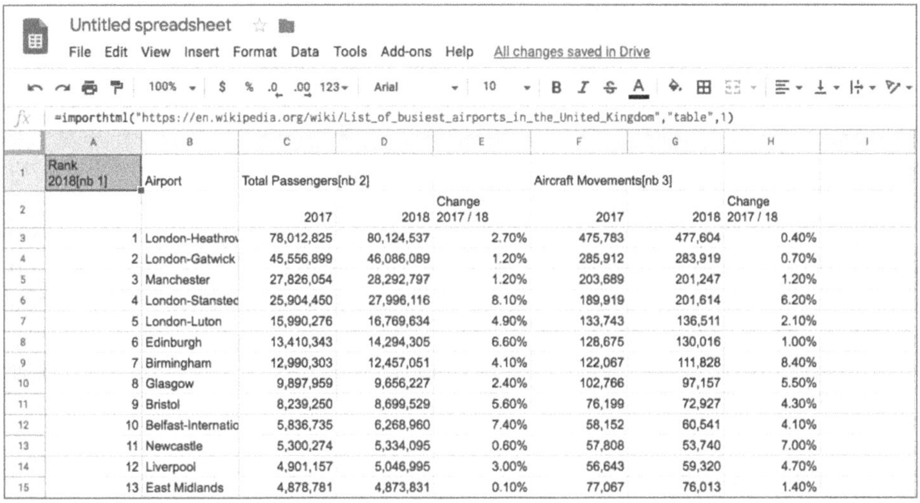

Figure 3.3: Spreadsheet of the busiest airports in the United Kingdom

To export the data to CSV format, click the File tab at the top of the spreadsheet, then click Download, and then save it to a comma-separated values file. This version of the data doesn't have the issues that the copy-and-paste version did. One thing to keep in mind is that the data is still text based; looking at the numbers, you can see they still have commas in their format.

While this method saves you a lot of time, there are still things you need to keep in mind. You'll have to do another round of cleaning to remove the commas on some of the number values. The best place to do that is in the spreadsheet itself and then export the data to CSV.

Using an API

When the whole Web 2.0 thing was being talked about in the early 2000s, the consensus was that everyone would have an API and we'd all acquire data from each other to power the Web. Personally, I'm not 100 percent convinced that happened. It did for some, but not many had the skills to acquire data in a machine-friendly and automated way.

An API is a set of routines supplied by a system or a website that lets you request and receive data or talk to a system directly. Most of the time you will have some sort of authority to talk to the service; this might be a token key or username and password combination, for example. Some APIs are public and don't require any sign-up, but those are rarer now because suppliers like to know who's calling, see what data you're taking, and know how often you're taking it.

Acquiring Weather Data

The website OpenWeather (`https://openweathermap.org`) has a full suite of APIs to retrieve weather information. There are various endpoints to get things like weather for a city or a three-day forecast and historical weather data. When you call the API service, you can specify the format you want the data to be in, whether that be JSON, CSV, or HTML.

Before you start, you will need to sign up at `openweathermap.org`. Once an account is created, you will need to take a copy of the API key that has been generated for you. Once your key is active, it can take a couple of hours; then you can try the examples.

For this example, I'm going to retrieve data from the API using three methods: the command line, Java, and then Clojure.

Using the Command Line

The `curl` command appears in most Linux distributions. There is a lot of power in this simple command that is worth investigating. For our uses now, it's quite simple because the weather API is a GET-based HTTP call. Using the `-o` flag, you can output the results to a file.

In the code repository for this book, there is a shell scripts directory and within the `ch03` folder a script that looks like the following:

```
#!/bin/bash

# Add your API key from openweathermaps.org
API_KEY=<<add your api key here>>

curl -o londonweather.json https://api.openweathermap.org/data/2.5/
weather?q=London\&APPID=${API_KEY}
```

You will need to add your API key from `openweathermap.org` to the shell script. When you run this from the command line, you'll see the following output:

```
$ ./openweather.sh
  % Total    % Received % Xferd  Average Speed Time    Time Time Current
                                 Dload  Upload   Total Spent    Left
Speed
100   457 100   457 0    0 3255 0 --:--:-- --:--:-- --:--:--  3264
```

When you open the `londonweather.json` file, you will see the JSON output.

```
{"coord":{"lon":-0.13,"lat":51.51},"weather":[{"id":800,"main":"Clear","d
escription":"clear sky","icon":"01n"}],"base":"stations","main":{"temp":
290.84,"pressure":1022,"humidity":68,"temp_min":288.15,"temp_max":294.15},
"visibility":10000,"wind":{"speed":3.6,"deg":90},"clouds":{"all":0},"dt":
1566593517,"sys":{"type":1,"id":1414,"message":0.009,"country":"GB","sunr
ise":1566536298,"sunset":1566587317},"timezone":3600,"id":2643743,"name":
"London","cod":200}
```

Using Java

The process of handling the URL and retrieving the content is all done by classes from the `java.io` and `java.net` packages. To convert the resulting string into a `JSONObject`, I'm using the `org.json` Java library.

When this code is executed, the first thing that happens is that the `readUrl` method is called with the URL to get data from. This is stored as a `String` object that is passed to the `stringToJSON` method to be converted into a JSON object (see Listing 3.1).

Listing 3.1: Using Java to Acquire Weather Data

```java
import org.json.JSONObject;
import java.io.BufferedReader;
import java.io.IOException;
import java.io.InputStreamReader;
import java.net.MalformedURLException;
import java.net.URL;
import java.net.URLConnection;

public class ReadURL {
    public String readUrl(String urlstring) {
        StringBuffer sb = new StringBuffer();
        try {
            URL url = new URL(urlstring);
            URLConnection urlConnection = url.openConnection();
            BufferedReader in = new BufferedReader(new
InputStreamReader(urlConnection.getInputStream()));
            String inputLine;
            while ((inputLine = in.readLine()) != null)
                sb.append(inputLine);
            in.close();
        } catch (MalformedURLException e) {
        } catch (IOException e) {
        }
        return sb.toString();
    }

    public JSONObject stringToJSON(String rawjson) {
        return new JSONObject(rawjson);
    }
```

```
public static void main(String[] args) throws Exception {
    String apikey = "Add your key here.....";
    ReadURL r = new ReadURL();
    String rawstring =r.readUrl("https://api.openweathermap.org/
data/2.5/weather?q=London&APPID=" + apikey);
    JSONObject j = r.stringToJSON(rawstring);
    System.out.println(j.toString());
}
}
```

Using Clojure

The Clojure language takes the power of the JVM but provides a functional and far more concise method of retrieving data. The `slurp` function can read in a file or a URL, and using the additional `clojure.data.json` library, you have a simple three-line function to read and convert JSON data from an API call.

```
(ns ch03.core
  (:require [clojure.data.json :as json])
  (:gen-class))

(def baseurl "https://api.openweathermap.org/data/2.5/weather?q=London&
APPID=")
(def apikey "Add your key here....")

(defn get-json []
  (let [rawstring (slurp (str baseurl apikey))]
    (json/read-str rawstring :key-fn keyword)))
```

It's worth noting that the `:key-fn` option is using the `keyword` function to convert JSON keys to map key identifiers that are used with Clojure.

```
ch03.core> (get-json)
{:coord {:lon -0.13, :lat 51.51}, :timezone 3600, :cod 200, :name
"London", :dt 1566597508, :wind {:speed 3.1, :deg 90}, :id 2643743,
:weather [{:id 800, :main "Clear", :description "clear sky", :icon
"01n"}], :clouds {:all 0}, :sys {:type 1, :id 1414, :message 0.0099,
:country "GB", :sunrise 1566536298, :sunset 1566587317}, :base
"stations", :main {:temp 289.73, :pressure 1022, :humidity 77, :temp_min
287.04, :temp_max 293.15}, :visibility 10000}
ch03.core>
```

Migrating Data

Acquiring data is one part of the equation; migrating and transforming it will also be requested at some point. For some jobs, writing a small program or script to import/export data would be fine, but as the volumes grow and the demands from stakeholders get more complex, we need to start looking at alternative tools.

Embulk is an open source bulk loading tool. It provides a number of plugins to read, write, and transform data. For example, if you wanted to read a directory of CSV files, transform them to JSON, and write them to AWS S3, that can be done with Embulk with a single configuration file. If you are using the OpenJDK, then it uses version 8 without any issues.

Installing Embulk

Embulk works on Linux, Mac, and Windows platforms. To install it on Linux and macOS, you will need to open a terminal window and execute the following four commands:

```
$curl --create-dirs -o ~/.embulk/bin/embulk -L "https://dl.embulk.org/
embulk-latest.jar"
$chmod +x ~/.embulk/bin/embulk
$echo 'export PATH="$HOME/.embulk/bin:$PATH"' >> ~/.bashrc
$source ~/.bashrc
```

Once it's installed, you can run Embulk from the command line as you would any other application.

Using the Quick Run

Embulk has a feature that will attempt to guess the schema of incoming data. In the `data/ch03/embulkdata` directory, you will see a CSV file generated from `http://www.fakenamegenerator.com`, which is a free service that generates test user data. Also in the same directory is the configuration file `simpleconfig.yml`.

The configuration file has an input step (`in:`) and an output step (`out:`).

```
in:
  type: file
  path_prefix: '/path/to/repo/./embulkdata/sample_'
out:
  type: stdout
```

When you execute Embulk, it will attempt to parse the CSV file and work out an input schema for you. Using the `-o` option, it will write the output YAML to a file.

```
$embulk guess ./embulkscripts/sampledata/simpleconfig.yml \
-o config.yml
```

If you take a look at the output file, you'll see that Embulk has now populated things like the delimiter type, whether to skip header lines and a representation of the schema.

```
in:
  type: file
  path_prefix: /path/to/repo/./embulkdata/sample_
  parser:
    charset: UTF-8
    newline: LF
    type: csv
    delimiter: ','
    quote: '"'
    escape: '"'
    trim_if_not_quoted: false
    skip_header_lines: 1
    allow_extra_columns: false
    allow_optional_columns: false
    columns:
    - {name: Number, type: long}
    - {name: Title, type: string}
    - {name: GivenName, type: string}
    - {name: MiddleInitial, type: string}
    - {name: Surname, type: string}
    - {name: City, type: string}
    - {name: ZipCode, type: string}
    - {name: Country, type: string}
    - {name: EmailAddress, type: string}
    - {name: Username, type: string}
    - {name: Age, type: long}
    - {name: Occupation, type: string}
    - {name: Company, type: string}
    - {name: GUID, type: string}
    - {name: Latitude, type: double}
    - {name: Longitude, type: double}
out: {type: stdout}
```

Installing Plugins

The core Embulk engine doesn't know the input and output types of the data it's working with; it's just coordinating the job that's being executed. Plugins are where the power of Embulk lies. For a full list of the plugins available, visit the www.embulk.org website.

Plugin installation is done from the command line. Use the following commands to either install a plugin or list the installed plugins on your machine:

```
$embulk gem install <embulk-plugin-name>
$embulk gem list
```

Now you know how to install plugins, I will cover two scenarios that commonly happen: migrating file-based data to a database and converting data from one type to another.

Migrating Files to Database

You've been asked to migrate some online review stats from a file dump in CSV format and migrate them to MySQL. While I appreciate it's easy to migrate a single file to MySQL database with the `mysqlimport` command, when there are many files in a directory, a more managed approach is required.

The schema for the MySQL database is in the same directory as the configuration. To install it, assuming you have MySQL installed (it will also be used in Chapter 12, "Machine Learning Streaming with Kafka"), run the following command to create the database:

```
$ mysqladmin -u root -p<yourpassword> create embulktest
```

Then import the schema.

```
$ mysql -u root -p<yourpassword> embulktest < schema.sql
```

The next job is to install the MySQL plugin from the Embulk repository. From the command line, run the following Embulk command:

```
$ embulk gem install embulk-output-mysql
2019-01-01 01:01:01.000 +0100: Embulk v0.9.17
Gem plugin path is: /home/jason/.embulk/lib/gems
Fetching: embulk-output-mysql-0.8.2.gem (100%)
Successfully installed embulk-output-mysql-0.8.2
1 gem installed
```

I'm using the simple config principle that I used in the previous example; I'm going to let Embulk do the work for me. This time, however, I've crafted the required output element with the information about the MySQL database and username and password information.

```
in:
  type: file
  path_prefix: '/path/to/repo/./embulkdata/file_to_db/output'
out:
  type: mysql
  host: localhost
  user: root
  password: xxxxx
  port: 3306
  table: scenario1
  database: embulktest
  mode: insert
```

When I run the guess function on Embulk, it will generate the `config.yml` as shown earlier, keeping the output element intact and updating the input element with the new information it's learned from the CSV file.

```
in:
  type: file
  path_prefix: /home/jason/./work/embulkscripts/sampledata/scenario1/
output
  parser:
    charset: UTF-8
    newline: CRLF
    type: csv
    delimiter: ','
    quote: '"'
    escape: '"'
    trim_if_not_quoted: false
    skip_header_lines: 1
    allow_extra_columns: false
    allow_optional_columns: false
    columns:
    - {name: userid, type: long}
    - {name: itemid, type: long}
    - {name: rating, type: double}
    - {name: timestamp, type: long}
out: {type: mysql, host: localhost, user: root, password: admin, port:
3307, table: scenario1,
  database: embulktest, mode: insert}
```

The final step is to run Embulk and apply the configuration. This will take data in the directory and insert it into the database.

```
$ embulk run config.yml
```

There will be a lot of message output while the job runs. Once it has completed, open up your MySQL database and then do a quick check.

```
$ mysql -u root -p<yourpassword> embulktest

mysql> select * from scenario1 limit 10;
+--------+--------+--------+------------+
| userid | itemid | rating | timestamp  |
+--------+--------+--------+------------+
|    548 |      5 |      3 |  857405447 |
|    292 |   1721 |    4.5 | 1140051202 |
|     73 |   3706 |    4.5 | 1464750953 |
|    378 |  95873 |    3.5 | 1443294223 |
|    165 |   1393 |      5 | 1111612302 |
|    553 |  59369 |      3 | 1423010662 |
|    104 |  42738 |    3.5 | 1446674082 |
|    283 |   6296 |      3 | 1115170015 |
|    548 |    544 |      3 |  857407872 |
|    353 |   1220 |      3 | 1157420794 |
+--------+--------+--------+------------+
10 rows in set (0.00 sec)
```

Bulk Converting CSV to JSON

One common request is converting data from one type to another. In this final example, I'll use Embulk to convert a CSV file to JSON. While it seems trivial, it can be done in code. What I'm doing is thinking forward to when the volumes of data are too big for single programs to handle.

The first thing to do is install the filter plugin, which will transform the data to JSON.

```
$ embulk gem install embulk-filter-to_json
2019-01-01 01:01:01.000 +0100: Embulk v0.9.17
```

In the csv_to_json example directory, you will see a data.csv file with scoring data. This is what will be converted to JSON. The same directory also has the configuration file for Embulk.

```
in:
  type: file
  path_prefix: data.csv
  parser:
    type: csv
    charset: UTF-8
    newline: CRLF
    null_string: 'NULL'
    skip_header_lines: 1
    comment_line_marker: '#'
    columns:
      - {name: time,  type: timestamp, format: "%Y-%m-%d"}
      - {name: id, type: long}
      - {name: name, type: string}
      - {name: score, type: double}
filters:
  - type: to_json
    column:
      name: test
      type: string
    skip_if_null: [id]
    default_timezone: Asia/Tokyo
out:
  type: stdout
```

Remember that the filter is not a CSV-to-JSON conversion; it's transforming to JSON anything that's passed in the process stream. When this example is run, the CSV data is passed through the input and then into the filter, and the resulting JSON output is sent to the console through the standard output channel.

```
$embulk run config.yml
```

Here's the sample output from my job execution. Note how any erroneous lines are skipped from the filter.

```
2019-08-24 10:42:04.983 +0100 [INFO] (0001:transaction): Loading files
[data.csv]
2019-08-24 10:42:05.180 +0100 [INFO] (0001:transaction): Using local
thread executor with max_threads=8 / output tasks 4 = input tasks 1 * 4
2019-08-24 10:42:05.198 +0100 [INFO] (0001:transaction): {done:  0 / 1,
running: 0}
2019-08-24 10:42:05.495 +0100 [WARN] (0014:task-0000): Skipped line
/home/jason/work/embulkscripts/sampledata/scenario3/data.csv:100
(org.embulk.spi.time.TimestampParseException: text is null or empty
string.): ,,,9170
{"score":1370.0,"name":"Vqjht6YEUBsMPXmoW1iOGFROZF27pBzz0TUkOKeDXEY","t
ime":"2015-07-13 09:00:00.000000000 +0900","id":0}
{"score":3962.0,"name":"VmjbjAA0tOoSEPv_vKAGMtD_0aXZji0abGe7_
VXHmUQ","time":"2015-07-13 09:00:00.000000000 +0900","id":1}
{"score":7323.0,"name":"C40P5H1WcBx-aWFDJCI8th6QPEI2DOUgupt_
gB8UutE","time":"2015-07-13 09:00:00.000000000 +0900","id":2}
{"score":5905.0,"name":"Prr0_u_T1ts4myUofBorOJFpCYcOTLOmNBMuRmKIPJU","t
ime":"2015-07-13 09:00:00.000000000 +0900","id":3}
{"score":8378.0,"name":"AEGIhHVW5cV6Xlb62uvx3TVl3kmh3Do8AvvtLDS7MDw","t
ime":"2015-07-13 09:00:00.000000000 +0900","id":4}
{"score":275.0,"name":"eupqWLrnCHr_1UaX4dUInLRxx5Q_cyQ4t0oSJBcw0MA","t
ime":"2015-07-13 09:00:00.000000000 +0900","id":5}
{"score":9303.0,"name":"BN8cQ47EXRb_oCGOoN96bhBldoiyoCp5O_
vGHwg0XCg","time":"2015-07-13 09:00:00.000000000 +0900","id":6}
```

Summary

In this chapter, I outlined a few techniques for acquiring data, whether that be via page scraping, using Google Sheets to import table data, or using scripting languages to clean up files. If an API is available, then it makes sense to maximize the potential gains from it whenever you can.

When the volumes of data start to build, then it's worth using tools designed for the job instead of crafting your own. The open source Embulk application is an excellent example of what has been created in the open source world. You can leverage it to speed up and streamline your data acquisition and migration strategies.

Statistics, Linear Regression, and Randomness

After acquiring and cleaning our data, it's now time to focus our attention on some numbers. As a gentle introduction, it's a good idea to revisit some statistics and how they can be used. In addition, I'll cover standard deviation, Bayesian techniques, forms of linear regression, and the power of random numbers.

The code to accompany this chapter will be in both Java and Clojure and will show you how to use some libraries as well as how to code these algorithms yourself.

Working with a Basic Dataset

Before we dive into this chapter, we require some data to work from. I have prepared a dataset of 474 scores from the judging of a television program (more on this later). They're all integers and give us a nice introduction into statistics.

As the chapter progresses, we'll add to this dataset and do some prediction work.

Loading and Converting the Dataset

You can download the dataset from the GitHub repository. In the folder /data /ch04 there is a file called stats.txt. As we are dealing with the text of numbers, there are some tasks that are required before we can start any work. Let's look at the file first, shown here:

```
2
5
3
4
...
3
9
8
8
```

/data/ch04/stats.txt

While it appears that there are numbers on each line of the text file, they are still treated as text. If we were to use mathematical notation at this point, our list of numbers would look like this:

```
{2, 5, 3, 4,...3,  9, 8, 8}
```

Our first task is to convert the contents of each line of the text file and convert them to an integer type that our program can understand.

Loading Data with Clojure

Reading a text file in Clojure can be done in one command using slurp and taking the file path as an argument. Slurping the file will consume it all, so there's some modification to do. This is called *transforming*.

Currently the file is one long line of numbers and newlines.

```
2\n5\n3\n4\n...3\n9\n8\n8
```

The split command in the clojure.string library will split on a given regular expression. This will produce a collection of strings. The last thing to do is to map through each string and cast it to a double value. The double parsing is using a Java function, as Clojure is a JVM language. We can call Java with ease using Java Interop.

```clojure
(defn load-file [filepath]
  (map (fn [v] (Double/parseDouble v))
       (-> (slurp filepath)
           (s/split #"\n"))))
```

Loading Data with Java

The process is identical to the Clojure process, though in the Java language it's a little more involved in terms of code. Using the `BufferedReader` and `FileReader` objects, a stream is created to read in the file. After iterating each line, it converts the value to an integer and adds it to the list.

Notice the use of the `Double` object to call the `parseDouble` method. It's the same method as used by the Clojure program.

```
package ch04;
import java.io.*;
import java.util.ArrayList;
import java.util.List;

public class LoadFileExample {
    public List<Double> loadFile(String filename) throws Exception {
        List<Double> numList = new ArrayList<Double>();
        File file = new File(filename);
        BufferedReader br = new BufferedReader(new FileReader(file));
        String s;
        while ((s = br.readLine()) != null) {
            numList.add(Double.parseDouble(s));
        }
        return numList;
    }

    public static void main(String[] args) throws Exception {
        List<Double> nums = new  LoadFileExample()
          .loadFile("/stats.txt");
        System.out.println(nums);
    }
}
```

Regardless of the method, the output is basically the same, a list of numbers.

```
[2, 5, 3, 4,.... 3, 9, 8, 8]
```

Assuming the resulting functions have been stored in a new object, then it's ready for use to get some summary statistics. In the following sections, we'll look at calculating some basis statistics with our vector of numbers.

Introducing Basic Statistics

I don't know why, but the mere mention of the word *statistics* can bring either a wide smile or a breakout of panic. There was a time I was in the former camp but transferred to the smiling camp. Regardless of how you feel about them, statistics are straightforward enough in code. I also include the mathematical notation for each of the summary statistic methods.

Covered in the section are the basic summary statistics: the sum, minimum and maximum, mean, mode, median, range, variance, and standard deviation.

Once again, I'll cover both Java and Clojure variations. With Clojure we have the bonus of having something called a REPL, which stands for "read, evaluate, print, loop," meaning you can type the commands out and get the results of code easily. Java sadly does not have this luxury in version 1.8, but there are services on the Internet that do provide REPL-like interfaces for Java if you want to experiment.

I will assume from this point on that you have the collection of scores in a value called numList.

Minimum and Maximum Values

Finding the minimum and maximum values of a list of numbers, while not seemingly groundbreaking in terms of stats or machine learning, is still worthwhile to know.

Mathematical Notation

It's perfectly fine to use the words *min* and *max*, but it's also acceptable to use an upper and lower arrow.

^ for the minimum value.

˅ for the maximum value.

Clojure

With Clojure we apply a function to the collection. This takes a function (in this instance either min or max) and uses the contents of the collection as an argument. If I were to pass directly to min or max, I would get the whole collection returned as it is classed as one argument.

```
(defn find-min-value [v]
  (apply min v))

(defn find-max-value [v]
  (apply max v))

;; Run on the REPL
ch04.core> (find-min-value numlist)
2.0
ch04.core> (find-max-value numlist)
10.0
```

Java

The `Collections` object will give you access to the methods `min` and `max` assuming that the input type is a collection. The `List<Integer>` type covered in the file loading example earlier in the chapter will work here.

```
Collections.min(numList);
Collections.max(numList);
```

Sum

The sum, or rather summation, is the addition of a sequence of numbers. The result is a single value of the total. The order of the numbers is not important in summation. For example, the summation of [1, 2, 3, 4] is the same as [3, 1, 4, 2].

Mathematical Notation

The mathematical notation for summation is the Greek letter sigma, which looks like a big *E*: Σ. The more we look at the algorithms used in machine learning, the more you'll see the adding up of a sequence or collection of numbers happens a lot.

Clojure

We're using the `apply` function against the collection again; the only change is the function that's being applied. The + is classed as a function.

```
(defn find-sum [v]
  (apply + v))

;; Run on the REPL
ch04.core> (find-sum numlist)
3113.0
```

Java

With Java, things require a little more thought, as we are dealing with a collection of objects. At this point, I could write a method to get the sum for me, iterating each value in the collection and adding to the accumulative total.

```
public int getSum(List<Integer> numList) {
    int total = 0;
    for(Integer i : numList){
        total += i.intValue();
    }
    return total;
}
```

An alternative would be to use the `Arrays` class and use the `stream()` method. Be aware that this method uses only primitive arrays as it's input, so you need to convert the `List` first.

```
int [] pNumList = list.stream()
                      .mapToInt(Integer::intValue)
                      .toArray();
int total = Arrays.stream(pNumList).sum();
```

Mean

The mean, or the average, is one of the first statistical methods you'll learn at school. When we say "the mean" or "the average," we are normally referencing the arithmetic mean. The mean gives us a good idea of where the middle is in a set of data.

However, there is a caveat to that: a nice smooth average is working with the assumption that the dataset is evenly distributed. If there are outliers within the dataset, then the average can be heavily distorted and incorrect. When there are outliers in the data, then it's wiser to use the median as a gauge.

Arithmetic Mean

To calculate the arithmetic mean, take the set of numbers and sum them. The last step is to divide that summed number by the number of items in the dataset.

$1 + 2 + 3 = 6$

$6 / 3 = 2$

Harmonic Mean

The harmonic mean is calculated differently. There are three steps to complete the calculation.

1. For each value, calculate the reciprocal value.
2. Find the average of the reciprocal values.
3. Calculate the reciprocal of the average.

$1/1 = 1, 1/2 = 0.5, 1/3 = 0.3333$

$1 + 0.5 + 0.3333 = 1.8333$

$3/1.8333 = 1.6366$

Geometric Mean

If the values in your dataset are widely different, then it's worth using the geometric mean to find the average. The calculation is made by multiplying the set of numbers and finding the *nth* root of the total. For example, if your set had two numbers in it, you'd square root the total; if it had three numbers, you would cube root; and so on.

The following are two examples, one with a set of three numbers and another with a set of six numbers.

$1 \times 2 \times 3 = 6$

$\sqrt[3]{6} = 1.81712$

The second example.

$1 \times 2 \times 3 \times 4 \times 5 \times 6 = 720$

$\sqrt[6]{720} = 2.9937$

The Relationship Between the Three Averages

There is a theory of mathematics called the *inequality of arithmetic and geometric means*, also known as the AM-GM inequality.

Within a list of numbers with no negative values, the arithmetic mean should be greater or equal to the geometric mean. The means of each type should be equal only when the values of the list are the same.

As a guide, the arithmetic mean should be equal or greater than the geometric mean, and the geometric mean should be equal or greater than the harmonic mean.

$AM \geq GM \geq HM$

In the examples for each of the means, we have the following outputs:

$2 \geq 1.81712 \geq 1.6366$

Now let's turn our attention to code and how to perform each of the mean types.

Clojure

For some of the Clojure code samples, I am using the `kixi.stats` library:

```
https://github.com/MastodonC/kixi.stats
```

You can easily run the examples from the REPL. Using the original dataset that was loaded in, you will get the following output:

```
(defn basic-arithmetic-mean [v]
  (/ (find-sum v) (count v)))

;; From the REPL
ch04.core> (basic-arithmetic-mean numlist)
6.567510548523207

(defn harmonic-mean [v]
  (transduce identity ks/harmonic-mean v))

;; From the REPL
ch04.core> (harmonic-mean numlist)
5.669668073229876

(defn geometric-mean [v]
  (transduce identity ks/geometric-mean v))

;; From the REPL
ch04.core> (geometric-mean (take 100 numlist))
5.917692496564965
```

The last example is slightly different from the others; I've used the `take` command to use the first 100 values from the dataset. The reason for this is that when all the values in the dataset are multiplied, the answer is infinity, meaning that the number has passed the maximum value of the data type. Using a subset of the full dataset reduces the chance of error.

Java

The Apache Commons Math library provides a useful set of summary statistics classes. Using the `StatUtils.mean` method will take a double primitive array and return the mean.

```
public double getMean(List<Double> nums) {
    double[] pNumList = nums.stream().mapToDouble(Double::doubleValue)
                            .toArray();
    return StatUtils.mean(pNumList);
}


public double getHarmonicMean(List<Double> nums) {
    double[] pNumList = nums.stream().mapToDouble(Double::doubleValue)
                            .toArray();
    double reciprocolTotal = 0.0;
    for(int i = 0 ; i < pNumList.length - 1 ; i++) {
        reciprocolTotal += 1/pNumList[i];
    }
```

```
   double harmonicMean = pNumList.length/reciprocolTotal;
   return harmonicMean;
}

public double getGeometricMean(List<Double> nums) {
   double[] pNumList = nums.stream().mapToDouble(Double::doubleValue)
                            .toArray();
   return StatUtils.geometricMean(pNumList);
}
```

Mode

To find the most commonly used number in the dataset, we use the mode.

Clojure

The `frequencies` command will tell you how many times a value has occurred in the dataset. This gives a map of value and frequency counts.

```
ch04.core> (frequencies numlist)
{2.0 15, 4.0 39, 8.0 91, 9.0 76, 5.0 43, 10.0 16, 3.0 36, 6.0 74, 7.0 84}
```

The next step is to use the `group-by` function to return another map, with the frequency value first and then a vector of the value/frequencies.

```
ch04.core> (group-by second (frequencies numlist))
{74 [[6.0 74]], 39 [[4.0 39]], 15 [[2.0 15]], 91 [[8.0 91]], 36 [[3.0 36]],
43 [[5.0 43]], 76 [[9.0 76]], 16 [[10.0 16]], 84 [[7.0 84]]}
```

Sorting that map gives you the frequencies in order. It's the last value we're interested in.

```
ch04.core> (last (sort (group-by second (frequencies numlist))))
[91 [[8.0 91]]]
```

We know value 8 has 91 occurrences; it's only the value 8 that we're wanting to return as the mode. Using the `map` function to find the first value of the second part of the vector (which is another vector, [8.0 91]), we get the result of the first element. That's the mode.

```
ch04.core> (map first (second (last (sort (group-by second (frequencies
numlist))))))
(8.0)
```

That can be wrapped up in a function; you can see this in the full code listing.

```
(defn find-mode [v]
  (map first (second
              (last
               (sort
                (group-by second
                          (frequencies v)))))))
```

Java

Use the `StatUtils.mode` method in Apache Commons Math to get the mode of a double primitive array. Notice it returns a double primitive array.

```
public double[] getMode(List<Double> nums) {
    double[] pNumList = nums.stream().mapToDouble(Double::doubleValue)
.toArray();
    return StatUtils.mode(pNumList);
}
```

Median

To find the middle number of the dataset, you use the median. Finding the median number involves listing the dataset in ascending order and finding the middle number.

If the total number of values in the dataset is odd, then the middle number is going to be a value from the dataset. On the other hand, if the dataset has an even set of values, then the average of the middle two numbers of the dataset is used.

Clojure

The `kixi.stats` library takes in a collection and will return the median.

```
(defn find-median [v]
  (transduce identity ks/median v))
```

Java

Using the `DescriptionStatistics` class, the `getPercentile` method will give the median from a collection. You will have to iterate the collection and add the double value to the instance of the class with the `addValue` method.

```
public double getMeadian(List<Double> nums) {
    double[] pNumList = nums.stream().mapToDouble(Double::doubleValue).
toArray();
```

```
DescriptiveStatistics ds = new DescriptiveStatistics();
for(int i = 0; i < pNumList.length -1 ; i++ ) {
    ds.addValue(pNumList[i]);
}
return ds.getPercentile(50);
}
```

Range

The range of the dataset is calculated by taking the minimum value of the set from the maximum value. So, for example, the dataset looks like this:

```
[2,2,3,4,5,7,7]
```

Then the range is $7 - 2 = 5$.

Clojure

You've seen the functions to find the minimum and the maximum values of the collection. Taking one away from the other will give you the range.

```
(defn find-range [numlist]
  (- (find-max-value numlist) (find-min-value numlist)))
```

Java

The same goes for the Java implementation. The methods for minimum and the maximum have already been established; it's just a case of reusing them.

```
public double getRange(List<Double> nums) {
    return (getMaxValue(nums) - getMinValue(nums));
}
```

Interquartile Ranges

As already discussed, if a dataset has outliers, the arithmetic mean will not be the centered average you are looking for. It's best using either the harmonic or geometric mean. The range gives a complete spread of the data, start to end. The interquartile range gives you the bulk of the values, also known as "the middle 50."

Subtracting the third quartile of the dataset from the first quartile will give you the interquartile range.

Clojure

The `kixi.stats` library has a function for the interquartile range.

```clojure
(defn interquartile-range [v]
  (transduce identity ks/iqr v))
```

Java

In the same way as finding the median, using the `DescriptiveStatistics` class will give you the interquartile range by subtracting the last quarter from the first quarter of the dataset.

```java
public double getIQR(List<Double> nums) {
    double[] pNumList = nums.stream().mapToDouble(Double::doubleValue)
                            .toArray();
    DescriptiveStatistics ds = new DescriptiveStatistics();
    for(int i = 0; i < pNumList.length -1 ; i++ ) {
        ds.addValue(pNumList[i]);
    }
    return ds.getPercentile(75) - ds.getPercentile(25);
}
```

Variance

The variance will give you the spread of the dataset. If you have a variance of zero, then all the values of the dataset are the same. There is a process to working out the variance of a dataset.

1. Work out the mean of the dataset.
2. For each number in the dataset, subtract the mean and then square the result.
3. Calculate the average of the squared differences.

Clojure

The variance can be found in the dataset with the `kixi.stats` library.

```clojure
(defn find-variance [numlist]
  (transduce identity ks/variance numlist))
```

Java

The `SummaryStatistics` class has a `getVariance` method. As with other examples, you will have to add values into the instance of the class with the `addValue` method.

```
public double getVariance(List<Double> nums){
    double[] pNumList = nums.stream().mapToDouble(Double::doubleValue)
                            .toArray();
    SummaryStatistics ss = new SummaryStatistics();
    for(int i = 0; i < pNumList.length -1 ; i++ ) {
        ss.addValue(pNumList[i]);
    }
    return ss.getVariance();
}
```

Standard Deviation

The standard deviation (sometimes called SD) is a number that tells us how the values for a dataset are spread out from the mean. If the standard deviation is low, then that means that most of the numbers in the dataset are close to the average. A large standard deviation will show that the numbers in the set are more spread out from the average.

The majority of the working out for the standard deviation is done by calculating the variance. The missing step is to square root the variance of the dataset.

The values that lie in the distribution can be calculated once you have the standard deviation. Called the empirical rule (or the 68-95-99.7 rule), it will tell you that 68 percent of the values will lie within two standard deviations to the mean, 95 percent within three and 99.7 percent within four.

Clojure

Standard deviation can be calculated with kixi.stats.

```
(defn find-standard-deviation [v]
  (transduce identity ks/standard-deviation v))
```

Java

The SummaryStatistics class supports standard deviation.

```
public double getStandardDeviation(List<Double> nums) {
    double[] pNumList = nums.stream().mapToDouble(Double::doubleValue)
                            .toArray();
    SummaryStatistics ss = new SummaryStatistics();
    for(int i = 0; i < pNumList.length -1 ; i++ ) {
        ss.addValue(pNumList[i]);
    }
    return ss.getStandardDeviation();
}
```

Using Simple Linear Regression

While linear regression is not a machine learning algorithm, it is classed as a statistical method. Regardless, being able to predict a value from historical data is a worthwhile skill to have at your disposal. Simple linear regression plots an independent variable (the predictor) against a dependent variable (criterion variable).

A good example uses the two commonly used temperature scales, Fahrenheit and Celsius, because there's a relationship between the two. It's illustrated with the following regression equation:

Fahrenheit = 1.8x + 32

Say we have a temperature reading of 28 Celsius. To find the Fahrenheit reading, we multiply 28 by 1.8 and add 32. The answer is 82.4f.

You can generate your own linear regression calculations easily either by using a spreadsheet or by using a library. In this example, we're going to use the comma-separated value file called `ch4slr.csv` and generate a simple linear regression by using an application and writing some code.

The data is comprised of two sets of scores from a competition. With the scores of the first judge, is it possible to reliably predict the scores of the second judge? We can find out by using simple linear regression.

Using Your Spreadsheet

No one that I'm aware of sits down and writes things out on paper that often. This is even more true when you have a lot of data, as we do with our score data. To impress your friends at dinner parties and other social gatherings, you can show them that you can do simple linear regression on a spreadsheet.

Using Excel

Within the graph functions of Excel, there are tools to enable linear regression. For this example, I'm using Microsoft Excel Office 365 edition. The same functionality exists in Libre Office and Open Office, and you can also work out simple linear regression in Google Sheets.

Loading the CSV Data

Start Excel, and the opening home screen will give you the option to create a new file or open an existing one. Click the Open button on the left.

Find the file `ch4slr.csv` and open it into Excel. This is just a two-column file representing two judges' scores from a competition (see Figure 4.1).

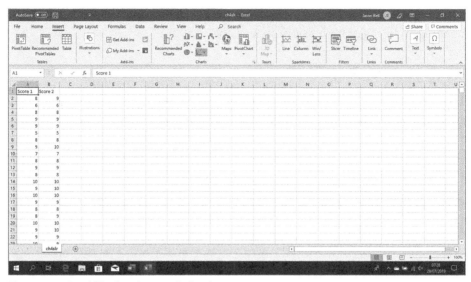

Figure 4.1: Excel file showing two judges' scores

Creating a Scatter Plot

The next step is to create a simple scatter plot graph. Select all the numbers in both columns and click Insert at the top. The top section of Excel will display a new set of icons; look for the Graph section, and you will see a scatter plot diagram. Clicking this will open a dialog box with scatter plot options.

Choose the Scatter option, which is the basic plot (see Figure 4.2).

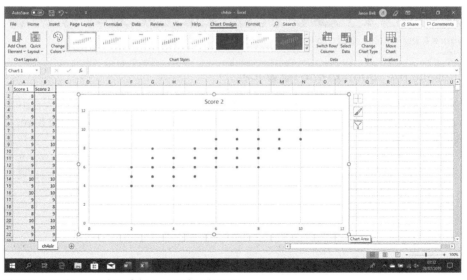

Figure 4.2: Scatter plot of the two judges' scores

The values of the CSV file will be displayed within the plot. There's little meaning in terms of regression, so let's add that in.

Showing the Trendline

First, I'd like to see a trendline to show where the data lies relative to the slope. Click the displayed scatter plot, and the options in the top menu will change. Click Add Chart Element, and a drop-down menu will appear. Select Trendline; then move your mouse across to the new menu and select Linear (see Figure 4.3).

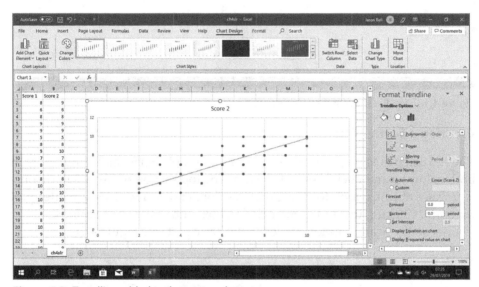

Figure 4.3: Trendline added to the scatter plot

Showing the Equation and R2 Value

Next up is the R2 value. As before, click Add Chart Element and select "Trendline. This time use the bottom option, More Trendline Options. This will bring a panel on the right side of the spreadsheet.

Scrolling down to the bottom of the panel, you will see three checkbox items. Click "Display Equation on chart" and "Display R-squared value on chart." The R2 value and the equation will appear on your chart (see Figure 4.4).

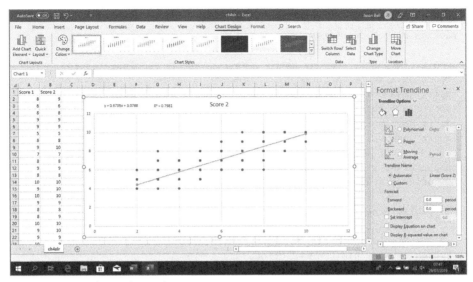

Figure 4.4: R2 value and equation

Making a Prediction

At this point you can use a calculator to make a prediction. Looking at the graph, I can see this equation:

y = 0.6735x + 3.0788

Assuming I want to predict what the judge's score will be if I rate a 6 in the competition, I can find out with the following equation:

Judge's score = (my score * 0.6735) + 3.0788

Or:

Judge's score = (6 * 0.6735) + 3.0788 = 7.1198

Rounding down, I get the score of 7.

Writing a Program

There comes a time when you will want to progress past a spreadsheet. This might be because there's so much data to process, for example.

When using Java, the Apache Commons Math library has an implementation of simple linear regression. The process is straightforward. The first step is to load the text file and add each comma pair into a collection (an `ArrayList` in this case). Using the `addData` method, the double values for both scores are passed in; the string to primitive double data type conversion happens during this step. The code for this is shown in Listing 4.1.

Listing 4.1: Using AddData for Simple Linear Regression

```java
package mlbook.chapter4.slr;

import org.apache.commons.math3.stat.regression.SimpleRegression;
import java.io.*;
import java.util.ArrayList;
import java.util.List;
import java.util.Random;
import java.util.UUID;

public class LinearRegressionBuilder {
    private static String path = "/path/to/ch4slr.csv";

    public LinearRegressionBuilder() {
        List<String> lines = loadData(path);
        SimpleRegression sr = getLinearRegressionModel(lines);
        System.out.println(runPredictions(sr, 40));
    }

    private SimpleRegression getLinearRegressionModel(List<String>
lines) {
        SimpleRegression sr = new SimpleRegression();
        for(String s : lines) {
            String[] ssplit = s.split(",");
            double x = Double.parseDouble(ssplit[0]);
            double y = Double.parseDouble(ssplit[1]);
            sr.addData(x,y);
        }
        return sr;
    }

    private String runPredictions(SimpleRegression sr, int runs) {
        StringBuilder sb = new StringBuilder();
        // Display the intercept of the regression
        sb.append("Intercept: " + sr.getIntercept());
        sb.append("\n");
        // Display the slope of the regression.
        sb.append("Slope: " + sr.getSlope());
        sb.append("\n");

        sb.append("\n");
        sb.append("");
        Random r = new Random();
        for (int i = 0 ; i < runs ; i++) {
            int rn = r.nextInt(10);
            sb.append("Input score: " + rn + " prediction: " +
                Math.round(sr.predict(rn)));
            sb.append("\n");
        }
        return sb.toString();
    }
```

```
    private List<String> loadData (String filename) {
        List<String> lines = new ArrayList<String>();
        try {
            FileReader f = new FileReader(filename);
            BufferedReader br;
            br = new BufferedReader(f);
            String line = "";
            while ((line = br.readLine()) != null) {
                lines.add(line);
            }
        } catch (FileNotFoundException e) {
            System.out.println("File not found.");
        } catch (IOException e) {
            System.out.println("Error reading file");
        }

        return lines;
    }

    public static void main(String[] args) {
        LinearRegressionBuilder dlr = new LinearRegressionBuilder();
    }
}
```

Running the program in Listing 4.1 will give different responses as the input scores are based on a random number. It will look something like this:

```
Intercept: 3.031026812343159
Slope: 0.6769332768870359
Running random predictions......
Input score: 4 prediction: 6
Input score: 5 prediction: 6
Input score: 2 prediction: 4
Input score: 5 prediction: 6
Input score: 3 prediction: 5
Input score: 8 prediction: 8
Input score: 4 prediction: 6
Input score: 9 prediction: 9
Input score: 8 prediction: 8
Input score: 3 prediction: 5
```

Embracing Randomness

It's not always essential for you to have data at hand to do any work. Random numbers can bring up some interesting experiments and code. In this section, we're going to look at two aspects of using random numbers. First we'll look at finding Pi using some basic math and Monte Carlo methods; second we'll look at random walks.

Finding Pi with Random Numbers

The Monte Carlo method is the concept of emulating a random process. When the process is repeated many times, it will give rise to the approximation of some mathematical quantity of interest. So, in theory with enough random darts thrown at a circle, you should be able to find the number of Pi.

Figure 4.5 shows our square.

Figure 4.5: Initial drawing of a square

Now draw a circle within the square (see Figure 4.6).

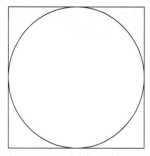

Figure 4.6: Circle within a square

Placing enough random data in the square will give you darts that are in the square, and some of them will be within the circle (see Figure 4.7). These are the darts that we're really interested in.

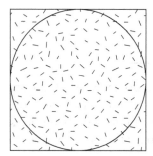

Figure 4.7: Random darts within the circle and the square

These are random throws. You might throw 10 times; you might throw 1 million times. At the end of the dart throws, you count the number of darts within the circle, divide that by the number of throws (10 million, 1 million, and so on), and then multiply it by 4.

π = 4 × (darts within the circle / total points)

The more throws we do, the better chance we get of finding a number near Pi. This is the law of large numbers at work. It's a classic computer science problem, and I'm going to solve it by writing a program in Clojure to do the simulation for us.

Using Monte Carlo Pi in Clojure

I'm going to create a function that simulates a single dart throw. I want to break down my Clojure code into as many simple functions as possible. This makes testing and bug finding far easier.

```
(defn throw-dart []
  {:x (calc-position 0)
   :y (calc-position 0)})
```

What I'm creating is a map with an x,y coordinate with a 0,0 center point and then passing the coordinate for x and y through another function to calculate the position (`calc-position`).

```
(def side-of-square 2)
(defn calc-position [v]
  (* (/ (+ v side-of-square) 2) (+ (- 1) (* 2 (Math/random)))))
```

The `calc-position` function takes the value of either x and y and applies the calculation. This is somewhere *-side-of-square*/2 and *+side-of-square*/2 around the center point.

Running this function in a REPL, we can see the x or y position.

```
mathematical.programming.examples.montecarlo> (calc-position 0)
0.4298901518005238
```

Is the Dart Within the Circle?

Now I have an x,y position as a map, `{:x some random throw value :y some random throw value}`, and I want to confirm that the throw is within the circle.

Using the `side-of-square` value again (hence it's a `def`), I can figure out if the dart hits within the circle. I'll pass the map with x,y coords in and take the square root of the added squared coordinates.

```
(defn is-within-circle [m]
  (let [distance-from-center (Math/sqrt (+ (Math/pow (:x m) 2)
                                           (Math/pow (:y m) 2)))]
    (< distance-from-center (/ side-of-square 2)))))
```

This function will return `true` or `false`. If I check this in the REPL, it looks like this:

```
mathematical.programming.examples.montecarlo> (throw-dart)
{:x 0.22535085231582297, :y 0.04203583357796781}
mathematical.programming.examples.montecarlo> (is-within-circle *1)
true
```

Now Throw Lots of Darts!

So far, there are functions to simulate a dart throw and confirm it's within the circle. Now I need to repeat this process as many times as required.

I'm creating two functions: `compute-pi-throwing-dart` to run a desired number of throws and `throw-range` to do the actual working to find the number of true hits in the circle.

```
(defn throw-range [throws]
  (filter (fn [t] (is-within-circle (throw-dart))) (range 0 throws)))

(defn compute-pi-throwing-dart [throws]
  (double (* 4 (/ (count (throw-range throws)) throws)))))
```

The `throw-range` function executes the `throw-dart` function, and `is-within-circle` evaluates the map to see whether the value is either true or false. The `filter` functions will return a list of true values. So, for example, if out of 10 throws the first, third, and fifth are within the circle, I'll get (1,3,5) as the result from the function.

Calling the function `compute-pi-throwing-dart` sets all this into motion. As I said at the beginning, taking the number of darts in the circle and dividing that by the number of throws taken and multiplying that by 4 should give a number close to Pi.

The more throws you do, the closer it should get.

Via the REPL, there is proof of an emergent behavior. The value of Pi comes from the large number of throws we did at the dart board.

The last thing I'll do is build a function to run the simulation.

```
(defn run-simulation [iter]
  (map (fn [i]
    (let [throws (long (Math/pow 10 i))]
      (compute-pi-throwing-dart throws))) (range 0 iter)))
```

If I run four simulations, I'll get 1, 10, 100, and 1,000 throws computed. These are then returned as a list. If I run nine simulations (which can take some time depending on the machine you're using) in the REPL, I get the following:

```
mathematical.programming.examples.montecarlo> (run-simulation 9)
(0.0 3.6 3.28 3.128 3.1176 3.1428 3.142932 3.1425368 3.14173752)
```

That's a nice approximation. Pi is 3.14159265, so getting the Monte Carlo method to compute Pi by random evaluations is good. Here is the final code listing:

```
(ns ch04.montecarlo)

(def side-of-square 2)

(defn calc-position [v]
  (* (/ (+ v side-of-square) 2) (+ (- 1) (* 2 (Math/random)))))

(defn throw-dart []
  {:x (calc-position 0)
   :y (calc-position 0)})

(defn is-within-circle [m]
  (let [distance-from-center (Math/sqrt (+ (Math/pow (:x m) 2)
                                           (Math/pow (:y m) 2)))]
    (< distance-from-center (/ side-of-square 2))))

(defn throw-range [throws]
  (filter (fn [t] (is-within-circle (throw-dart))) (range 0 throws)))

(defn compute-pi-throwing-dart [throws]
  (double (* 4 (/ (count (throw-range throws)) throws))))

(defn run-simulation [iter]
  (map (fn [i]
         (let [throws (long (Math/pow 10 i))]
           (compute-pi-throwing-dart throws))) (range 0 iter)))

mathematical.programming.examples.montecarlo> (run-simulation 3)
(4.0 3.6 3.28)
mathematical.programming.examples.montecarlo> (run-simulation 9)
(0.0 4.0 3.28 3.056 3.1392 3.146 3.139848 3.1414404 3.14128264)
mathematical.programming.examples.montecarlo>
```

Summary

Mathematics underpins everything that is done within machine learning. This chapter acts as a reminder to some of the basic summary statistics, building on this knowledge to produce techniques like linear regression, standard deviation, and Monte Carlo methods.

Adding simple programming functions with Java and Clojure, you now have a suite of tools at your disposal whenever you need them. Don't forget there are times when a spreadsheet wins.

Working with Decision Trees

Do not be deceived by the decision tree; at first glance it might look like a simple concept, but within the simplicity lies its power. This chapter shows you how decision trees work. The examples use Weka to create a working decision tree that will also create the Java code for you.

The Basics of Decision Trees

The aim of any decision tree is to create a workable model that will predict the value of a target variable based on the set of input variables. This section explains where decision trees are used along with some of the advantages and limitations of decision trees. In this section, you also find out how a decision tree is calculated manually so you can see the math involved.

Uses for Decision Trees

Think about how you select different options within an automated telephone call. The options are essentially decisions that are being made for you to get to the desired department. These decision trees are used effectively in many industry areas.

Financial institutions use decision trees. One of the fundamental use cases is in option pricing, where a binary-like decision tree is used to predict the price of an option in either a bull or bear market.

Marketers use decision trees to establish customers by type and predict whether a customer will buy a specific type of product.

In the medical field, decision tree models have been designed to diagnose blood infections or even predict heart attack outcomes in chest pain patients. Variables in the decision tree include diagnosis, treatment, and patient data.

The gaming industry now uses multiple decision trees in movement recognition and facial recognition. The Microsoft Kinect platform uses this method to track body movement. The Kinect team used one million images and trained three trees. Within one day and using a 1,000-core cluster, the decision trees were classifying specific body parts across the screen.

Advantages of Decision Trees

There are some good reasons to use decision trees. For one thing, they are easy to read. After a model is generated, it's easy to report to others regarding how the tree works. Also, with decision trees you can handle numerical or categorized information. Later, this chapter demonstrates how to manually work through an algorithm with category values; the example walk-through uses numerical data.

In terms of data preparation, there's little to do. As long as the data is formalized in something like comma-separated variables, then you can create a working model. This also makes it easy to validate the model using various tests. With decision trees you use white-box testing—meaning the internal workings can be observed but not changed; you can view the steps that are being used when the tree is being modeled.

Decision trees perform well with reasonable amounts of computing power. If you have a large set of data, then decision tree learning will handle it well.

Limitations of Decision Trees

With every set of advantages there's usually a set of disadvantages sitting in the background. One of the main issues of decision trees is that they can create overly complex models, depending on the data presented in the training set. To avoid the machine learning algorithm's over-fitting the data, it's sometimes worth reviewing the training data and pruning the values to categories, which will produce a more refined and better-tuned model.

Some of the decision tree concepts can be hard to learn because the model cannot express them easily. This shortcoming sometimes results in a larger-than-normal model. You might be required to change the model or look at different methods of machine learning.

Different Algorithm Types

Over the years, there have been various algorithms developed for decision tree analysis. Some of the more common ones are listed here.

ID3

The *ID3* (Iterative Dichotomiser 3) algorithm was invented by Ross Quinlan to create trees from datasets. By calculating the entropy for every attribute in the dataset, this could be split into subsets based on the minimum entropy value. After the set had a decision tree node created, all that was required was to recursively go through the remaining attributes in the set.

ID3 uses the method of information gain—the measure of difference in entropy before and after an attribute is split—to decide on the root node (the node with the highest information gain).

ID3 suffered from over-fitting on training data, and the algorithm was better suited to smaller trees than large ones. The ID3 algorithm is used less these days in favor of the C4.5 algorithm, which is outlined next.

C4.5

Quinlan came back for an encore with the C4.5 algorithm. It's also based on the information gain method, but it enables the trees to be used for classification. This is a widely used algorithm in that many users run in Weka with the open source Java version of C4.5, the J48 algorithm.

There are notable improvements in C4.5 over the original ID3 algorithm. With the ability to work on continuous attributes, the C4.5 method will calculate a threshold point for the split to occur. For example, with a list of values like the following:

```
85,80,83,70,68,65,64,72,69,75,75,72,81,71
```

C4.5 will work out a split point for the attribute (a) and give a simple decision criterion of:

```
a <= 80 or a > 80
```

C4.5 has the ability to work despite missing attribute values. The missing values are marked with a question mark (?). The gain and entropy calculations are simply skipped when there is no data available.

Trees created with C4.5 are pruned after creation; the algorithm will revisit the nodes and decide if a node is contributing to the result in the tree. If it isn't, then it's replaced with a leaf node.

CHAID

The *CHAID* (Chi-squared Automatic Interaction Detection) technique was developed by Gordon V. Kass in 1980. The main use of it was within marketing, but it was also used within medical and psychiatric research.

MARS

For numerical data, it might be worth investigating the *MARS* (multivariate adaptive regression splines) algorithm. You might see this as an open source alternative called "Earth," as MARS is trademarked by Salford Systems.

How Decision Trees Work

Every tree is comprised of nodes. Each node is associated with one of the input variables. The edges coming from that node are the total possible values of that node. A leaf represents the value based on the values given from the input variable in the path running from the root node to the leaf. Because a picture paints a thousand words, see Figure 5.1 for an example.

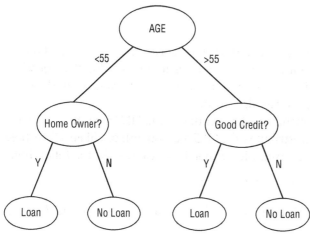

Figure 5.1: A decision tree

Decision trees always start with a root node and end on a leaf. Notice that the trees don't converge at any point; they split their way out as the nodes are processed.

Figure 5.1 shows a decision tree that classifies a loan decision. The root node is "Age" and has two branches that come from it, whether the customer is younger or older than 55.

The age of the client determines what happens next. If the person is younger than 55, then the tree prompts you to find out if he or she is a student. If the client is older than 55, then you are prompted to check his or her credit rating.

With this type of machine learning, you are using supervised learning to deduce the optimal method to make a prediction; what I mean by "supervised learning" is that you give the classifier data with the outcomes. The real question is, "What's the best node to start with as the root node?" The next section examines how that calculation is done.

Building a Decision Tree

Decision trees are built around the basic concept of this algorithm.

- Check the model for the base cases.
- Iterate through all the attributes (`attr`).
- Get the normalized information gain from splitting on `attr`.
- Let `best_attr` be the attribute with the highest information gain.
- Create a decision node that splits on the `best_attr` attribute.
- Work on the sublists that are obtained by splitting on `best_attr` and add those nodes as child nodes.

That's the basic outline of what happens when you build a decision tree. Depending on the algorithm type, like the ones previously mentioned, there might be subtle differences in the way things are done.

Manually Walking Through an Example

If you are interested in the basic mechanics of how the algorithm works and want to follow along, this section walks through the basics of calculating entropy and information gain. If you want to get to the hands-on part of the chapter, then you can skip this section.

The method of using information gain based on pre- and post-attribute entropy is the key method used within the ID3 and C4.5 algorithms. As these are the commonly used algorithms, this section concentrates on that basic method of finding out how the decision tree is built.

With machine learning–based decision trees, you can get the algorithm to do all the work for you. It will figure out which is the best node to use as the root node. This requires finding out the purity of each node. Consider Table 5.1, which includes only true/false values, of some user purchases through an e-commerce store.

Table 5.1: Users' Purchase History

	HAS CREDIT ACCOUNT?	READ REVIEWS?	PREVIOUS CUSTOMER?	DID PURCHASE?
User A	N	Y	Y	Y
User B	Y	Y	Y	Y
User C	N	N	Y	N
User D	Y	N	N	Y
User E	Y	Y	Y	Y

There are four nodes in the table:

- Does the customer have an account?
- Did the customer read previous product reviews?
- Is the customer a returning customer?
- Did the customer purchase the product?

At the start of calculating the decision tree there is no awareness of the node that will give the best result. You're looking for the node that can best predict the outcome. This requires some calculation. Enter entropy.

Calculating Entropy

Entropy is a measure of uncertainty and is measured in bits and comes as a number between zero and 1 (entropy bits are not the same bits as used in computing terminology). Basically, you are looking for the unpredictability in a random variable.

There are two entropy types to work out; the first is entropy on a single attribute, and the second is entropy for two attributes. With this example, I'll use the Had Credit Account as our target attribute for our working example. By the end of the exercise, we will know the next node to use in the decision tree.

The single attribute for the Has Credit Account attribute has two outcomes (Yes or No).

HAS CREDIT ACCOUNT	
Yes	No
3	2

The entropy calculation looks as follows:

$$
\begin{aligned}
\text{Entropy(Has Credit Account)} &= \text{Entropy}(2,3) \\
&= \text{Entropy}(0.4, 0.6) \\
&= -(0.4 \log_2 0.4) - (0.6 \log_2 0.6) \\
&= 0.97095
\end{aligned}
$$

When two attributes are applied, the values are mapped as in the following table:

		HAS CREDIT ACCOUNT		
		Y	N	
Reads Reviews	Y	2	1	3
	N	1	1	2

With two attributes, we end up with the following entropy calculation:

$$\begin{aligned}
&\text{Entropy}\left(\text{Has Credit Account, Reads Reviews}\right)\\
&= P(Y) \times E(2,1) + P(N) \times E(1,1)\\
&= (2/5) \times 0.971 + (1/5) \times 1\\
&= 0.5884
\end{aligned}$$

So, you have two gains: one before the split (Has Credit Account, which is 0.97095) and one after the split (Has Credit Account/Reads Reviews, which is 0.5884).

You're nearly done on this attribute. The next step is to calculate the information gain.

Information Gain

When you know the gain before and after the split in the attribute, you can calculate the information gain. With the attribute to see if the customer has a credit account, your calculation will be the following:

$$\begin{aligned}
\text{InformationGain} &= \text{Gain(before the split)} - \text{Gain(after the split}\\
&\quad \text{with Reads Reviews)}\\
&= 0.97095 - 0.5884\\
&= 0.38255
\end{aligned}$$

So, the information gain on the Has Credit Account attribute is 0.38255.

Rinse and Repeat

The previous two sections covered the calculation of information gain for one attribute, Has Credit Account. You need to work on the other two attributes to find their information gain. With the values of information gain for all the attributes, you can now make a decision on which node to start with in the tree.

ATTRIBUTE	INFORMATION GAIN
Reads Reviews	0.38255
Did Make Purchase	0.48672
Is Previous Customer	0.37095

Now things are becoming clearer; assuming the Has Credit Account is the root node, the Did Make Purchase attribute has the highest information gain and therefore, should be the next deciding node in the tree.

The order of information gain determines where the node will appear in the decision tree model. The node with the highest gain becomes the root node. Working out all the gains on each attribute and their information gains will mathematically give you the tree.

As the decision tree algorithm is calculated iteratively, smaller trees can be worked out easily. Though it's easy to assume that with several hundreds of instances no one really wants to work it all out by hand. At that point it's time to bring in software to help.

Decision Trees in Weka

In this section, you'll use the Weka data-mining tool to work through some training data of the optimum sales of Lady Gaga's CDs depending on specific factors within the store. I explain the factors in question as you walk though that data. While I appreciate the world has leaned towards streaming and downloading music, people still do buy CDs.

The Requirement

The requirement is to create a model that will be able to predict a customer sale on Lady Gaga CDs depending on the CDs' placement within the store. You've been given some data by the record store about where the product was placed, whether it was at eye level or not, and whether the customer actually purchased the CD or put it back on the shelf.

The client wants to be able to run other sets of data through the model to determine how sales of a product will fare.

Working through this methodically, you need to do the following:

1. Run through the training data supplied and turn it into a definition file for Weka.

2. Use the Weka workbench to build the decision tree for you and plot an output graph.

3. Export some generated Java code with the new decision tree classifier.

4. Test the code against some test data.

5. Think about future iterations of the classifier.

It feels like there's a lot to do, but after you get into the routine, it's quite simple to accomplish with the tools at hand. First look at the training data.

Training Data

Before anything else happens, you need some training data. The client has given you some in a `.csv` file, but it would be nice to formalize this. This is what you received:

```
Placement,prominence, pricing, eye_level, customer_purchase
end_rack,85,85,FALSE,yes
end_rack,80,90,TRUE,yes
cd_spec,83,86,FALSE,no
std_rack,70,96,FALSE,no
std_rack,68,80,FALSE,no
std_rack,65,70,TRUE,yes
cd_spec,64,65,TRUE,yes
end_rack,72,95,FALSE,yes
end_rack,69,70,FALSE,yes
std_rack,75,80,FALSE,no
end_rack,75,70,TRUE,no
cd_spec,72,90,TRUE,no
cd_spec,81,75,FALSE,yes
std_rack,71,91,TRUE,yes
```

Weka saves the file as an `.arff` file to set up the attributes and let you give it some data from which to train. The `.arff` file is a text file that outlines the data model you are going to use:

```
@relation ladygaga

@attribute placement {end_rack, cd_spec, std_rack}
@attribute prominence numeric
@attribute pricing numeric
@attribute eye_level {TRUE, FALSE}
@attribute customer_purchase {yes, no}

@data
end_rack,85,85,FALSE,yes
end_rack,80,90,TRUE,yes
cd_spec,83,86,FALSE,no
std_rack,70,96,FALSE,no
std_rack,68,80,FALSE,no
std_rack,65,70,TRUE,yes
cd_spec,64,65,TRUE,yes
end_rack,72,95,FALSE,yes
end_rack,69,70,FALSE,no
std_rack,75,80,FALSE,no
end_rack,75,70,TRUE,no
cd_spec,72,90,TRUE,no
cd_spec,81,75,FALSE,yes
std_rack,71,91,TRUE,yes
```

The data file has a few elements to it, so let's look through it one section at a time.

Relation

The @relation tag is the name of the dataset you are using. In this instance it's Lady Gaga's CDs, so I've called it ladygaga.

Attributes

Next, you have the attributes that are used within your data model. There are five attributes in this set that are the top line of raw CSV data that you received from the client.

Placement: What type of stand the CD is displayed on: an end rack, a special offer bucket, or a standard rack?

Prominence: What percentage of the CDs on display are Lady Gaga CDs?

Pricing: What percentage of the full price was the CD at the time of purchase? Rarely is a CD sold at full price, unless it is an old, back-catalog title.

Eye Level: Was the product displayed at eye-level position? The majority of sales will happen when a product is displayed at eye level.

Customer Purchase: What was the outcome? Did the customer purchase?

The Prominence and Pricing attributes are both numeric values. The other three are given the nominal values that are to be expected when the algorithm is being run. Placement has three: end_rack, cd_spec, or std_rack. The Eye Level attribute is either true or false, and the Customer Purchase attribute has two nominal values of either yes or no to show that the customer bought the product.

Data

Finally, you have the data. It's comma separated in the order of the attributes (Placement, Prominence, Pricing, Eye Level, and Customer Purchase). In this sample, you know the outcomes—whether a customer purchased or not; this model is about using regression to get your predictions in tune for new data coming in.

You can find all the code for this chapter on the book's companion website at

www.wiley.com/go/machinelearning2e

Using Weka to Create a Decision Tree

Now that you have your data model in place, you can get started. When you open the Weka program, you are presented with a small opening screen (see Figure 5.2) with four buttons: Explorer, Experimenter, KnowledgeFlow, and Simple CLI. Click the Explorer button.

Figure 5.2: The Weka GUI Chooser

When the Explorer opens, you will be confronted with another window with a number of sections and an array of buttons (see Figure 5.3). Don't worry if it all looks confusing right now; this walk-through takes you through it step by step.

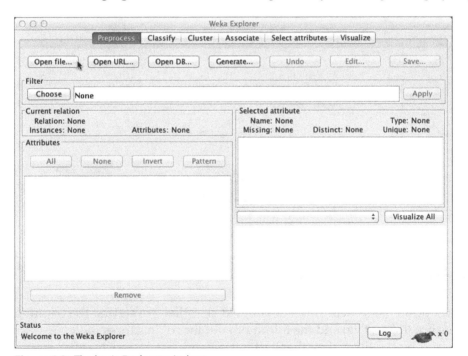

Figure 5.3: The basic Explorer window

Click the Open File button and select the data file called `ladygaga.arff`. Weka parses the data model and preprocesses the data. Within no time you're already getting information based on the preprocessing of the data model and the data.

The Select Attribute pane on the right side of the Explorer window in Figure 5.4 shows the three distinct nominal values of the `customer_pur-chase` attribute. Weka has also noticed that you have 14 instance rows and the five attributes.

After preprocessing comes classification. Click the Classify button in the top row of buttons. You're going to use the C4.5 classification algorithm; within Weka this is called the J48 algorithm. In the Classifier pane (see Figure 5.5), click the Choose button and select the J48 option under the Trees menu heading. The selection pane closes automatically, and you see that the name of the classifier has changed from the default `ZeroR` to `J48 -C 0.25 -M 2` (see Figure 5.5).

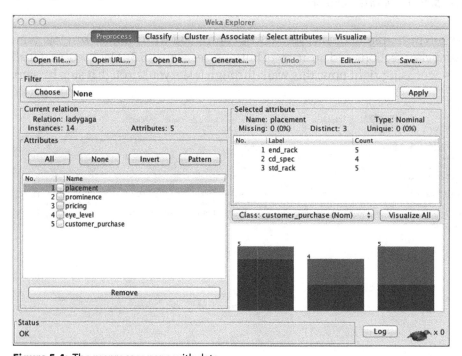

Figure 5.4: The preprocess pane with data

The option flags used in the default J48 classifier are setting the pruning confidence (the `-C` flag) and the minimum number of instances (`-M`).

To run the classifier, click the Start button and watch the Classifier output window (see Figure 5.6). You see the information on the run appear. The run information tells you about the scheme used and gives a run-down on the model on which Weka has worked.

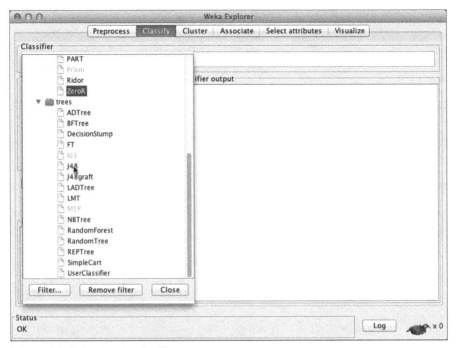

Figure 5.5: Selecting the classifier

Interesting data starts to emerge. The J48 pruned tree gives results on, in this case, the placement, as it has the highest information gain:

```
J48 pruned tree
------------------

placement = end_rack: yes (5.0/1.0)
placement = cd_spec
|    pricing <= 80: yes (2.0)
|    pricing > 80: no (2.0)
placement = std_rack
|    eye_level = TRUE: yes (2.0)
|    eye_level = FALSE: no (3.0)

Number of Leaves  :      5

Size of the tree :   8



Time taken to build model: 0 seconds
```

It appears that placing the product on the end rack is good for sales. For the special offer rack, it seems that pricing plays a part; if the product is too cheap customers, walk away. On the standard racks, the placement of the product is a factor for sales; it sells if it's at eye level.

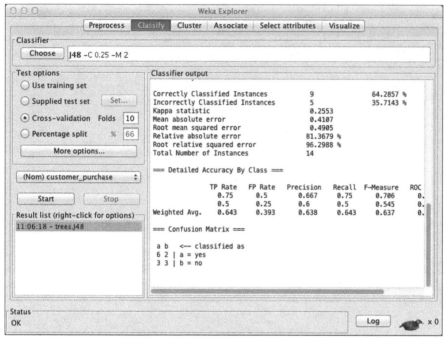

Figure 5.6: Classifier with output

Finally you want to plot the visualization of the tree for the management team to look at because pictures speak louder than words. On the Results List pane on the bottom left of the Explorer window you can see the time and algorithm that was run. Right-click (use Alt+click if you are using a macOS machine) and select the Visualize Tree option to see the tree in its visual representation, as shown in Figure 5.7.

It's usually at this point where everyone pats each other on the back and says, "Job well done," but you're not finished yet. You don't want to have to run the Weka Explorer every time you have data to run. What you want is some code that you can reuse.

Creating Java Code from the Classification

As mentioned in Chapter 2, there is no one tool that really fits all. Weka is excellent, but you want code that you can safely run in an existing codebase. Perhaps you want to hook your newly created classification to a Hadoop job, if the incoming volume of data was sufficient to do so.

With the existing classifier, click the More Options button, and a new window opens with the options for the current evaluator. (See Figure 5.8.)

The last option is to output to source code. By default, the class name will be WekaClassifier. It won't save your Java code, but it will output in the Classifier output window.

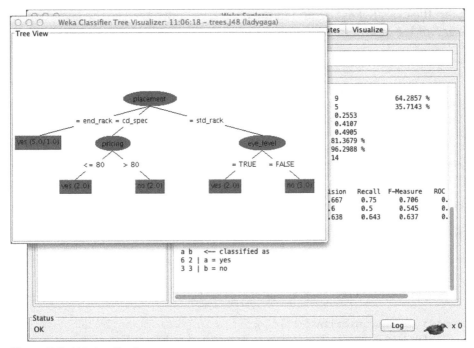

Figure 5.7: J48 visualization

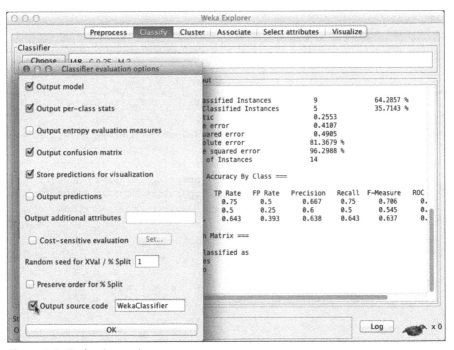

Figure 5.8: Evaluation options pane

Start the classifier again, and in the output window you see the Java code at the end of the output information.

```java
package mlbook.ch5.decisiontrees;

import weka.core.Attribute;
import weka.core.Capabilities;
import weka.core.Capabilities.Capability;
import weka.core.Instance;
import weka.core.Instances;
import weka.core.RevisionUtils;
import weka.classifiers.Classifier;

public class WekaWrapper
  extends Classifier {

  /**
   * Returns only the toString() method.
   *
   * @return a string describing the classifier
   */
  public String globalInfo() {
    return toString();
  }

  /**
   * Returns the capabilities of this classifier.
   *
   * @return the capabilities
   */
  public Capabilities getCapabilities() {
    weka.core.Capabilities result = new weka.core.Capabilities(this);

    result.enable(weka.core.Capabilities.Capability.NOMINAL_ATTRIBUTES);
    result.enable(weka.core.Capabilities.Capability.NUMERIC_ATTRIBUTES);
    result.enable(weka.core.Capabilities.Capability.DATE_ATTRIBUTES);
    result.enable(weka.core.Capabilities.Capability.MISSING_VALUES);
    result.enable(weka.core.Capabilities.Capability.NOMINAL_CLASS);
    result.enable(weka.core.Capabilities.Capability.MISSING_CLASS_
VALUES);

    result.setMinimumNumberInstances(0);

    return result;
  }

  /**
   * only checks the data against its capabilities.
   *
   * @param i the training data
   */
```

```java
public void buildClassifier(Instances i) throws Exception {
  // can classifier handle the data?
  getCapabilities().testWithFail(i);
}

/**
 * Classifies the given instance.
 *
 * @param i the instance to classify
 * @return the classification result
 */
public double classifyInstance(Instance i) throws Exception {
  Object[] s = new Object[i.numAttributes()];

  for (int j = 0; j < s.length; j++) {
    if (!i.isMissing(j)) {
      if (i.attribute(j).isNominal())
        s[j] = new String(i.stringValue(j));
      else if (i.attribute(j).isNumeric())
        s[j] = new Double(i.value(j));
    }
  }

  // set class value to missing
  s[i.classIndex()] = null;

  return WekaClassifier.classify(s);
}

/**
 * Returns the revision string.
 *
 * @return        the revision
 */
public String getRevision() {
  return RevisionUtils.extract("1.0");
}

/**
 * Returns only the classnames and what classifier it is based on.
 *
 * @return a short description
 */
public String toString() {
  return "Auto-generated classifier wrapper, based on weka.
classifiers.trees.J48 (generated with Weka 3.6.10).\n" + this.
getClass().getName() + "/WekaClassifier";
}

/**
 * Runs the classfier from commandline.
 *
```

```
    * @param args the commandline arguments
    */
  public static void main(String args[]) {
    runClassifier(new WekaWrapper(), args);
  }
}

class WekaClassifier {

  public static double classify(Object[] i)
      throws Exception {

    double p = Double.NaN;
    p = WekaClassifier.N32ec89882(i);
    return p;
  }
  static double N32ec89882(Object []i) {
    double p = Double.NaN;
    if (i[0] == null) {
      p = 0;
    } else if (i[0].equals("end_rack")) {
      p = 0;
    } else if (i[0].equals("cd_spec")) {
    p = WekaClassifier.N473959d63(i);
    } else if (i[0].equals("std_rack")) {
    p = WekaClassifier.N63915224(i);
    }
    return p;
  }
  static double N473959d63(Object []i) {
    double p = Double.NaN;
    if (i[2] == null) {
      p = 0;
    } else if (((Double) i[2]).doubleValue() <= 80.0) {
      p = 0;
    } else if (((Double) i[2]).doubleValue() > 80.0) {
      p = 1;
    }
    return p;
  }
  static double N63915224(Object []i) {
    double p = Double.NaN;
    if (i[3] == null) {
      p = 0;
    } else if (i[3].equals("TRUE")) {
      p = 0;
    } else if (i[3].equals("FALSE")) {
      p = 1;
    }
    return p;
  }
}
```

Open your text editor of choice and then copy and paste the Java code. Save the file as `WekaClassifier.java` (or the name of the class you specified in the options pane).

In the source code, there are actually two classes: a wrapper class that Weka generates and a main method from which to run. The core of the classifier is in the second class, `WekaClassifier`. This is basically a set of if/then statements based on the classified tree.

Testing the Classifier Code

Make a copy of the `.arff` file to test your coded classifier. Where the outcomes are yes or no, replace them with question marks (?). This means you want the classifier to work out the answer for you:

```
end_rack,85,85,FALSE,?
end_rack,80,90,TRUE,?
cd_spec,83,86,FALSE,?
std_rack,70,96,FALSE,?
std_rack,68,80,FALSE,?
std_rack,65,70,TRUE,?
cd_spec,64,65,TRUE,?
end_rack,72,95,FALSE,?
end_rack,69,70,FALSE,?
std_rack,75,80,FALSE,?
end_rack,75,70,TRUE,?
cd_spec,72,90,TRUE,?
cd_spec,81,75,FALSE,?
std_rack,71,91,TRUE,?
```

You need to write a new class to load in your test data and run each instance against the coded classifier.

```java
package chapter3;

import java.io.BufferedReader;
import java.io.FileReader;

import weka.core.Instances;

public class TestClassifier {
    public static void main(String[] args) {
        WekaWrapper ww = new WekaWrapper();
        try {
            Instances unlabeled = new Instances(new BufferedReader(
                    new FileReader("lg2.arff")));

            unlabeled.setClassIndex(unlabeled.numAttributes() - 1);
```

```
              for (int i = 0; i < unlabeled.numInstances(); i++) {
                  double clsLabel =
      ww.classifyInstance(unlabeled.instance(i));
                  System.out.println(clsLabel + " -> " +
      unlabeled.classAttribute().value((int) clsLabel));
              }
          } catch (Exception e) {
              e.printStackTrace();
          }
      }

  }
```

The instances are loaded in, and then the `for` loop iterates and uses the generated `classifyInstance()` method to get the scoring from the classifier. In this example, you're looking for the decision of whether a sale will happen.

Because the `classifyInstance()` returns the value as a double data type, you reference that against the class attribute array position. In this case, the `customer_purchase` attribute has two elements only, "yes" and "no." The first element in the array (0) points to "yes," and the second element (1) points to "no."

Running this example generates the following output:

```
0.0 -> yes
0.0 -> yes
1.0 -> no
1.0 -> no
1.0 -> no
0.0 -> yes
0.0 -> yes
0.0 -> yes
0.0 -> yes
1.0 -> no
0.0 -> yes
1.0 -> no
0.0 -> yes
0.0 -> yes
```

You could develop this basic code further to pull the required information from a database via Java Database Connectivity (JDBC) and then store the results again. You could even dump the results into a text file by making a copy of the instances first and updating them in the `for` loop.

```
Instances unlabeled = new Instances(new BufferedReader(
              new FileReader("lg2.arff")));
unlabeled.setClassIndex(unlabeled.numAttributes() - 1);
Instances trained = new Instances(unlabeled);

for (int i = 0; i < unlabeled.numInstances(); i++) {
    double clsLabel = ww.classifyInstance(unlabeled.instance(i));
```

```
    trained.instance(i).setClassValue(clsLabel);
    System.out.println(clsLabel + " -> " +
      unlabeled.classAttribute().value((int) clsLabel));
}
```

The changes required are labeled in bold. This would be useful if you were to output the changes of the instances to a text file, for example.

In terms of the actual work, you're done. You can deliver some solid code.

Thinking About Future Iterations

This chapter covered a lot of ground in a short space: putting an .arff file together to create a classifier and generating the Java code with Weka and testing it with more unclassified data.

The test data you had was small, which is fine for getting everything working. In the real world, though, you'd be processing much more data. The question is, how much data should you retain for training? As a guide, I use 10 percent of the total data as a starting point and work from there. It's also worth thinking about the seasonality of data, especially if you are working in retail. Creating models for certain seasonal periods can boost the information gain in your training sets.

Time waits for no one, and the same applies here. Data changes, trends change, and so do management decisions, and so on. It's important to keep the classifier up-to-date by means of running new test data and seeing if the model can improve.

Summary

You've seen how decision trees work and the different algorithm types that are available. At a hands-on level, you've worked on a full project to create a working classifier based on the C4.5 (J48, which is the Java open source implementation as used in Weka) algorithm to predict customer purchasing behavior on products determined by placement, prominence, and pricing. Although many people perceive decision trees as simple, do not underestimate their uses. They are easy to understand and don't need a huge amount of preparation. They are often useful regardless of whether you have category or numerical data.

Clustering

One of the more common machine learning strands you'll come across is clustering, mainly because it's very useful. For example, marketing companies love it because they can group customers into segments. This chapter describes the details of clustering and illustrates how clusters work and where they are used.

NOTE Please don't confuse machine learning clustering with clusters of machines in a networking sense; we're talking about something very different here.

What Is Clustering?

If you boil down all the definitions of clustering out there, you get "organizing a group of objects that share similar characteristics." It's classed as an unsupervised learning method, which means there's no prior training data from which to learn. In Figure 6.1 you see there are three distinct groupings of data; each one of those groups is a cluster.

The main aim is to find structure within a given set of data. Because there are a lot of algorithms to choose from, clustering casts a wide net. This is where experimentation comes in handy; which algorithm is the right choice? Sometimes you just need to put some code together and play with it. You'll do that shortly.

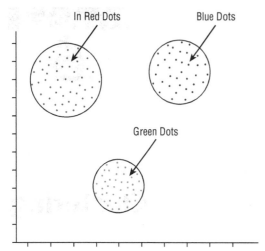

Figure 6.1: A graph representation of a cluster

Where Is Clustering Used?

Clustering is a widely used machine learning approach. Although it might seem simple, do not underestimate the importance of grouping multivariate data into refined groupings.

The Internet

Social media network analysis uses clustering to determine communities of users. With so many users on Facebook, for example, using these sorts of techniques can refine advertising so that certain ads go to specific groups of customers.

If you've ever searched on one of many mapping websites, you might have seen clustering at work when there are a lot of interest pins built up in a given location. Instead of showing all the pins, which would provide a bad user experience, clustering is used to define the group of pins within the given location.

Website logs and search results are often clustered to show more relevant search result groups. A number of companies are using clustering to refine search engine queries.

Business and Retail

Market research companies use clustering a lot. With surveys that contain many variables, a multivariate system like clustering gives marketers a better definition of groups of customers in relation to the answers given in the survey. This might be broken down by population, location, and previous buying habits, for example. With the clusters defined, the marketing companies can try to develop new products or think about testing products for certain clusters in the results.

Along with association rules learning (discussed in Chapter 7), clustering is also used for basket analysis. Certain auction sites use clustering because there is no defined stock number in the listings. It's easier to run clustering and group items and preferences than use association rules.

Law Enforcement

Crimes are logged with all the aspects of the felony listed. Police departments are running clustering and other machine learning algorithms to predict when and where future crimes will happen. The result of this might be that patrol cars are deployed to certain problem areas at certain times, or specialist help is sent to areas where certain sorts of crimes show high numbers.

Computing

With the rise of the "Internet of Things," we are now collecting more data from sensors than ever before. Clustering can be used to group the results of the sensors. For example, thinking of a temperature sensor, you might cluster date and time against the temperature. Another example would be motion detection; a number of passive infrared sensors could be generating data on movement within a location. Is a certain location a hotspot at a specific time? This information can be easily clustered and inspected.

Course work in the education sector, especially with the advent of large-scale learning online, can be clustered into student groups and results. For example, do certain clusters of students excel at courses compared to other students?

Clustering is used often in digital imaging. When large groups of images need to be segmented, it's usually a cluster algorithm that works on the set and defines the clusters. Algorithms can be trained to recognize faces, specific objects, or borders, for example.

Clustering Models

As previously mentioned, the goal of clustering is to segment data into specific groups. There are many different clustering algorithms for the simple fact that there is really only one common denominator among all clusters—that you're trying to find groups of objects.

For example, there are distribution models that use multivariate distributions for their modeling. Graph models can show cluster-like properties when the nodes start showing as small subsets connected with one main edge, as shown in Figure 6.2.

You can also approach simple clustering with groups in the same way you group in a structured query language.

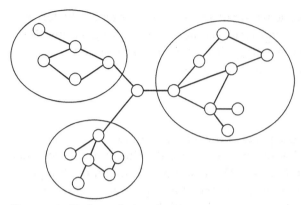

Figure 6.2: Nodes and edges as clusters

One of the more commonly used cluster models is the centroid model, which is where the k-means algorithm comes in. The *k-means algorithm* is basically vector quantization. This chapter concentrates on the k-means algorithm and creates a basis of the walk-through later in the chapter.

How the K-Means Works

If you have a group of objects, the idea of the k-means algorithm is to define a number of clusters. What's important is that it's up to you to define how many clusters you want. For example, say I have 1,000 objects and I want to find 4 clusters:

$$n(\text{objects}) = 1000$$
$$k(\text{clusters}) = 4$$

Each one of the clusters has a centroid (sometimes called the *mean*, hence the name *k-means*), a point where the distance of the objects will be calculated. The clusters are defined by an iterative process on the distances of the objects to calculate which are nearest to the centroid. This is all done unsupervised; you just have to let the algorithm do its processing and inspect the results. After the iterations have taken place to the point where the objects don't move to different centroids, then it's assumed that the k-means clustering is complete.

The following pseudocode describes what's happening:

```
calculate inital values for means m[1],m[2],m[3],m[4]

assign object to nearest center

while(there are changes in the mean position) {
    estimate the means to classify into clusters
    for(i in 1 to k) {
        m[i] = mean of the samples for cluster i
    }
}
```

Initialization

First, the algorithm must initialize by assigning a cluster to every observation made. The *random partition method* places the cluster points toward the center of the dataset. Another initialization method is the *Forgy method*, which spreads out the randomness of the initial location of the cluster.

After the initial cluster observations are assigned, you can look at the assignment and updating of the algorithm.

Assignments

Each observed object is assigned to the cluster to find out which cluster centroid it's assigned to; the algorithm uses a Euclidean distance measurement. The sum of squares is then calculated by squaring the Euclidean distances to each cluster centroid, and the one with the smallest value is the cluster that the object is assigned to.

Calculating the Euclidean distance is quite simple and requires only some entry-level math; if you can remember how to do Pythagoras' theorem, then you are already there.

Assume a basic grid of six positions on the X-axis (horizontal) and four positions on the Y-axis (vertical). The center point of my cluster is currently at 1,6, and the object is located at 3,1, as shown in Figure 6.3.

The distance is 3-1=2 on the vertical side and 6-1=5 on the horizontal axis. Using Pythagoras' theorem, the squared distance is

$$2 \wedge 2 + 5 \wedge 2 = 4 + 25 = 29$$

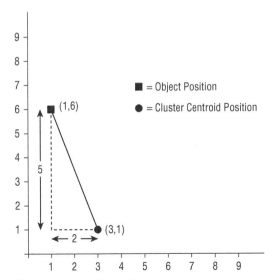

Figure 6.3: Euclidean distances

The square root of 29 is 5.38. This carries on for all the objects in the dataset that are assigned to the clusters.

Update

In the update step, you assign the object values to the cluster. The recalculation of the center points of the cluster (the centroids) is taken as the average of the values of the objects that are part of the cluster. This process carries on as a loop until the entities in each group no longer change.

The k-means algorithm is very effective, but it's not without its problems. It can take a few runs with the data to get a decent fit of clusters. When you choose too few, objects can easily spill into the incorrect cluster over a period of time during the processing.

Calculating the Number of Clusters in a Dataset

When presented with a dataset, it can be hard to define the number of clusters that you want to classify against. Sometimes this number will already be determined by a stakeholder. For example, in a marketing initiative you might have the following:

- Low-frequency, low-value customers
- Low-frequency, high-value customers
- High-frequency, low-value customers
- High-frequency, high-value customers

There are times when this information is not available, and you have to find a balance for making your decisions. There are a number of methods for calculating the optimum.

The Rule of Thumb Method

Nothing beats wetting your finger and sticking it in the air to see which way the wind is blowing. There's a simple calculation that is roughly the equivalent for clusters: the number of clusters (k) is equal to the square root of the number of objects divided by 2.

$$K = \sqrt{objects / 2}$$

If you have 250 objects, then half of that is 125, and the square root of 125 is 11.18—so, there are 11 clusters. This can obviously be tested and reapplied depending on how the trial runs go.

The Elbow Method

You can calculate the variance of the dataset as a percentage and plot against the number of clusters. There's a point at which the clusters are at an optimum—that point after which adding more clusters will not make a huge difference to the final classifications. You can see in Figure 6.4 how the elbow method shows the optimum number of clusters is four, which has classified 80 percent of the data.

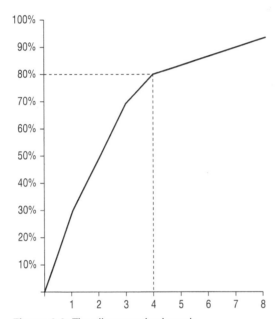

Figure 6.4: The elbow method graph

The Cross-Validation Method

By splitting the dataset into separate partitions, you can apply the analysis on the dataset and then on the remaining partitions. By averaging the results of the sum of squares, you can determine the number of clusters to use.

   Weka supports cross-validation with the `weka.clusterers.MakeDensity Based-Clusterer` class. This class is covered in more detail in the command-line-based walk-through later in the chapter.

The Silhouette Method

Peter J. Rousseeuw first described the *silhouette method* in 1986. It is a method for suggesting a way of validating where the objects lay within a cluster.

   For any object, you can calculate how similar an object is with another object within the same cluster. By calculating the averages of objects that connect to a cluster and then evaluating how dissimilar they are in relation to the other clusters, you can determine an average score. The main aim is to measure the grouping of the objects in the cluster—the lower the number the better. When comparing the averages for each cluster, they are expected to be similar. When silhouettes are very narrow and others are large, it might point to the fact that not enough clusters have been defined when the computation process began.

K-Means Clustering with Weka

The Weka machine learning application comes with an algorithm for processing k-means clusters, a class called SimpleKMeans. In this walk-through, you'll work with three approaches: one from the workbench application, one that works directly from the command line, and finally one that's a Java coded example.

The aim is to take some marketing data and use the k-means to generate some segmentation of the customers; this will potentially give the marketing department an increase in successful transactions in the long run.

Before you get into the three walk-throughs, you need to prepare some data with which to work.

Preparing the Data

The following Java code generates 75 instances of random numbers for two integer variables, x and y. You import the saved file (kmeansdata.csv) into Weka and also load it in programmatically.

```java
import java.io.BufferedWriter;
import java.io.FileWriter;
import java.io.IOException;
import java.util.Random;

public class DataSet3D {
    public static void main(String[] args) {
        Random r = new Random(System.nanoTime());
        try {
            BufferedWriter out =
new BufferedWriter(new FileWriter("kmeansdata.csv"));
            out.write("x,y\n");
            for(int count = 0; count < 75; count++) {
                out.write(r.nextInt(125) + "," + r.nextInt(150) + "\n");
            }
            out.close();
        } catch (IOException e) {
            e.printStackTrace();
        }

    }
}
```

The output looks something like this:

```
x,y
78,29
0,55
101,19
52,146
49,140
44,97
65,45
41,49
66,141
111,100
23,128
101,1
1,113
88,100
```

If you want to create more instances, you can do so by adjusting the number of iterations in the `for` loop of the code. I've set it to 75 as a starting point. The upper limit is really based on the amount of memory your computer has.

Have a look at the Weka workbench method first.

The Workbench Method

The *workbench method* uses the Weka user interface to load, cluster, and then visualize the data. There's no actual programming involved, but it's useful to see what Weka is doing before you progress on to the command line and coded samples.

Loading Data

The process for the workbench is similar to other examples you've run though. The first thing you need to do is load in the CSV data.

Click the Open File button and select the `kmeansdata.csv` file you created earlier. Ensure that the file format drop-down menu shows CSV and not ARFF (see Figure 6.5); otherwise, you won't be able to open the file.

When the data has loaded, the explorer shows various pieces of information. The Current Relation pane shows that there are two attributes and 75 instances. (See Figure 6.6.) The attribute information shows the two attributes: x and y.

On the right-hand panel, the Selected Attribute pane shows some statistics of the data that's been loaded, which includes the minimum and maximum values along with the mean and the standard deviation. Finally, there's a graph of the distribution of the values and the frequency of them.

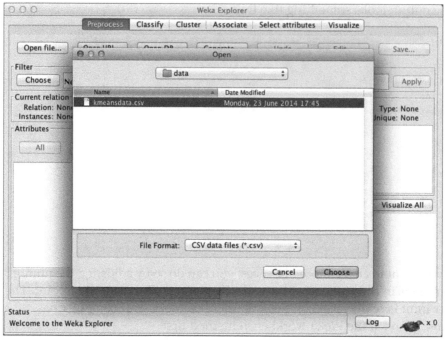

Figure 6.5: Loading CSV data into Weka

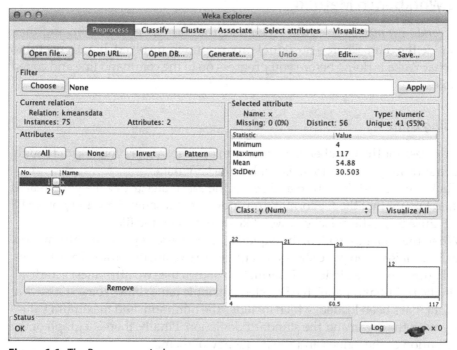

Figure 6.6: The Preprocess window

Clustering the Data

Click the Cluster tab at the top to select the clustering method. By default, the Weka clusterer uses the Simple EM method (expectation maximization). Clicking the Choose button displays a tree of other cluster algorithms that you can use. For this example, select SimpleKMeans and then click Close; finally, click Start to run the algorithm, as shown in Figure 6.7.

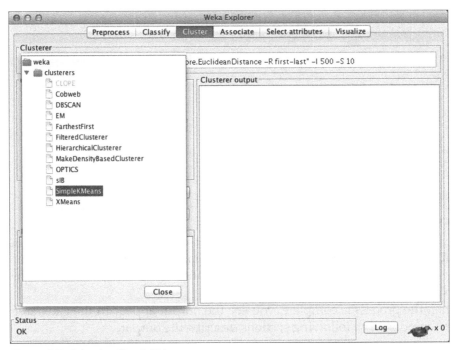

Figure 6.7: Selecting SimpleKMeans

The clusterer line in the Clusterer output pane has updated and shows the new cluster algorithm you'll be using:

```
SimpleKMeans -N 2 -A "weka.core.EuclideanDistance -R \
first -last" -1 500 -S 10
```

There are a few flags that need a little explanation.

- -N determines the number of clusters that the SimpleKMeans is going to create.

- -A is the distance function used. It defaults to Euclidean distance and uses the entire range of values as its range to act on (-R first -last).

- The -1 flag defines the number of iterations the k-means does to define the cluster.

- -S is a random number seed. It can be any value you want.

Clicking the line with all the command options shown next to the Choose button displays a pop-up window where you can alter the values. In the num-Clusters field, change the number from 2 to 4. (See Figure 6.8.) You're going to create four clusters with this data. Click OK to close the window.

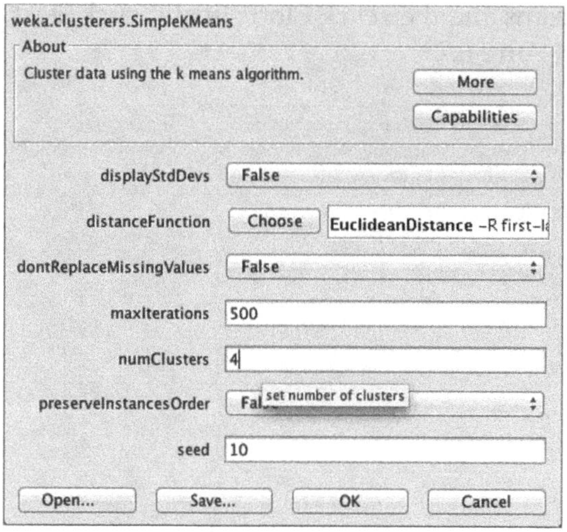

Figure 6.8: Changing the SimpleKMeans options

Start the clustering process by clicking the Start button. For long periods of processing, it's worth keeping an eye on the status in the bottom-left corner of the explorer.

When the process is complete, you see the output appear in the Clusterer Output window. The following sections describe the output.

Run Information

The first block of information gives the settings of the data and the algorithm selected.

```
Scheme:weka.clusterers.SimpleKMeans -N 4 -A
"weka.core.EuclideanDistance -R first-last" -I 500 -S 10
Relation:       td
Instances:      75
Attributes:     2
                x
                y
Test mode:evaluate on training data
```

K-Means

The k-means output shows the actual work that Weka did to reach the results. This includes the number of iterations that were performed on the data and the sum of squared errors.

```
Number of iterations: 3
Within cluster sum of squared errors: 0.8072960323968902
Missing values globally replaced with mean/mode

Cluster centroids:
                          Cluster#
Attribute     Full Data          0          1          2          3
                  (75)        (15)       (23)       (17)       (20)
===================================================================
x               54.88     68.9333     43.913    98.1765      20.15
y             92.0267        19.4   146.0435   114.8824      64.95
```

The cluster centroid data information is shown in relation to the instance data, showing the final value locations of the centroids for each cluster.

Clustered Instances

Finally, the percentage of the data within each cluster is shown. It gives you an idea of how the data is distributed.

```
Clustered Instances

0      15 ( 20%)
1      23 ( 31%)
2      17 ( 23%)
3      20 ( 27%)
```

Visualizing the Data

The last thing to do is look at the visualization of the clustering. In the explorer window, you see the result list on the bottom left. Right-clicking (or Alt+clicking on the Mac) the SimpleKMeans brings up the visualize window, as shown in Figure 6.9.

Each cluster has its own color scheme, and the plot shows values. Along the top of the visualization window are two, drop-down menus. To alter the plot, select a value from each drop-down menu. To show the x and y instance values, select the X: x(Num) and Y: y(Num) values, and you see the plot update.

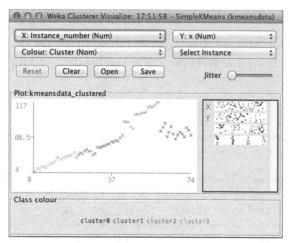

Figure 6.9: Visualize window

Using the workbench method gives you a quick route to some insight, and it is useful if you want to get a grasp on where the clusters lie. To start automating the process, you need to look at the other two approaches, starting with the command-line method.

The Command-Line Method

The *command-line method* is similar to the workbench method, but it gives you a little more flexibility in terms of running within cron jobs. This means that you can collate more data and rerun the analysis on a regular basis.

Converting CSV File to ARFF

Weka uses the .arff format to determine object types and have data ready for processing. The GUI enables you to import .csv files directly, but it's nice to have a tool that converts files for you.

You can use the CVSLoader class on the command line to convert .csv files to .arff. For example, the .csv file you created in the previous walk-through was kmeansdata.csv. Use the converter by running the following from the terminal command line:

```
java -cp weka.jar weka.core.converters.CSVLoader \
kmeansdata.csv > kmeansdata.arff
```

NOTE   If you're using the Windows operating system then you can omit the -cp weka.jar from the java command.

The command-line output might give some warnings about database drivers not being available, but there's no need to worry about that. The main thing is that in the .arff file you have the proper definition.

Do a quick inspection to see the following definition:

```
@relation kmeansdata

@attribute x numeric
@attribute y numeric

@data
```

The First Run

Test the command-line methods by starting with just running the SimpleKMeans class as- is on the .arff file. This is your training file:

```
java -cp /path/to/weka.jar weka.clusterers.SimpleKMeans \
  -t kmeandata.arff
```

The -t flag gives you the name of the training file that Weka is attempting to cluster. Running this as-is gives the following output:

```
kMeans
======

Number of iterations: 3
Within cluster sum of squared errors: 5.839457872519278
Missing values globally replaced with mean/mode

Cluster centroids:
                          Cluster#
Attribute     Full Data         0            1
                   (75)       (35)         (40)
==============================================
x                 54.88    41.0571       66.975
y               92.0267    45.4286        132.8




=== Clustering stats for training data ===

Clustered Instances
0       35 ( 47%)
1       40 ( 53%)
```

That works okay, but it's only two clusters, and there's a good chance there are more.

Refining the Optimum Clusters

Working out the optimum with the rule of thumb method described earlier in the chapter is easy enough. You can find out the number of object instances using the UNIX wc command.

```
wc kmeansdata.csv
      75        76      494 kmeansdata.csv
```

There are 74 lines (excluding the top line, which gives you the data labels x and y). A quick calculation of 75 divided by 2 results in 37.5, and the square root of that is 6.12.

By altering the command line, you can add that target cluster number (using the -N flag) along with a random seed number to work off (using the -S flag).

```
java -cp /path/to/weka.jar weka.clusterers.SimpleKMeans \
-t kmeansdata.arff -N 6 -S 42
```

The output this time gives the same output but with more clusters.

```
kMeans
======

Number of iterations: 3
Within cluster sum of squared errors: 0.523849925862059
Missing values globally replaced with mean/mode

Cluster centroids:
                          Cluster#
Attribute  Full Data     0          1          2        3        4        5
            (75)        (10)       (12)       (15)      (5)     (10)     (23)
==============================================================================
x          54.88        11.8      105.0833   68.9333   81.6    28.5     43.913
y          92.0267      65.9      118.3333   19.4      106.6   64       146.0435



=== Clustering stats for training data ===

Clustered Instances
0        10 ( 13%)
1        12 ( 16%)
2        15 ( 20%)
3         5 (  7%)
4        10 ( 13%)
5        23 ( 31%)
```

Now, you have six clusters with a good definition of objects going into their specific clusters. The only problem is that you don't know to which cluster the objects belong.

Name That Cluster

From the command line, the -p flag tells Weka to display the assignment of each row's cluster instance. To use this feature, you have to instruct Weka which data attribute to use for each row. From the command line, use the following:

```
java -cp /path/to/weka.jar weka.clusterers.SimpleKMeans \
-t kmeansdata.arff -N 6 -S 42 -p 0
```

The -p 0 flag tells Weka to display the row and cluster based on the row number of the data. When this is run, you see the following output to the console:

```
0  0
1  0
2  0
3  0
4  0
5  0
6  0
7  0
8  0
9  0
10 4
11 4
12 4
13 4
14 4
15 4
16 4
17 4
18 4
19 4
```

All that's being shown is the row number and the numeric identifier of the cluster to which the row belongs. If you set the -p flag to 1 or 2, then you'd get the value of the x or y positions, respectively.

With these workbench and command-line examples, you should now have a good idea how all this is put together. Take a look at the Java coded method using the Weka application programming interface (API). It demonstrates how you can integrate these clustering methods into your own server-side projects.

The Coded Method

The workbench works well, and the command line suits development needs better when scheduled jobs need to be done. But, for ultimate flexibility, there's nothing better than coding your own program using the API to get things done.

The same core Weka classes are used by the workbench, command line, and Java coded examples. It's a case of figuring out how the data is loaded and the clustering done with the options you've used in previous examples.

Creating a k-means cluster in Java is actually a fairly trivial matter; it's a matter of knowing what elements go where. You are going to create a simple Java program to complete what you've done in the previous two walk-throughs. I'll be using Eclipse.

Create the Project

Select File ➪ New ➪ Java Project and call it WekaCluster, as shown in Figure 6.10.

Figure 6.10: Eclipse New Java Project dialog box

There's only one library to install; that is the weka.jar file. On macOS machines, Weka is usually installed within the Applications directory. The location on Windows machines varies depending on the specific operating system.

With the WekaCluster project selected, click File ➪ Properties and look for Java Build Path. Then, click the Libraries tab. Add the external jar file by clicking Add External JARs. In the File dialog box, find the weka.jar file, as shown in Figure 6.11.

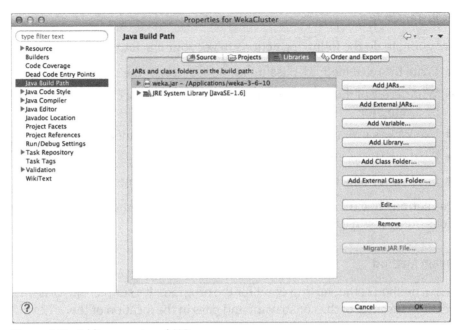

Figure 6.11: Adding an external JAR

The last thing to do is create a new class called `WekaCluster.java` (use File ➪ New ➪ Class); see Figure 6.12 for what this should look like.

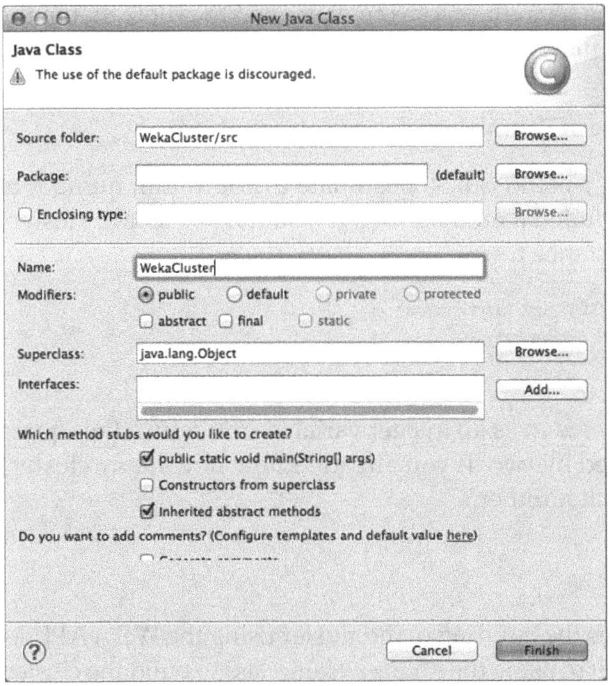

Figure 6.12: Creating a new class file

The Cluster Code

You're going to do the following actions to get your cluster working:

- Write the `main` method, passing in the location of the `.arff` file.
- Write a rule of thumb routine to advise the number of clusters you should be aiming for.
- Use Weka's `SimpleKMeans` class to build the cluster model.
- Print out the location of the centroids of the cluster.
- Print out to which cluster each instance object belongs.

This sounds like a lot of work, but it's actually quite simple. The following sections break down each individual step.

The main Method

The `main` method is the starting point for the program. It's simple—just one line to create an instance of the constructor and pass in the location of the `.arff` file.

```
public static void main(String[] args) {
    // Pass the arff location and the number of clusters we want
    WekaCluster wc = new WekaCluster("/path/to/kmeandata.arff");
}
```

Because you're passing the filepath as a string, you need to reflect that in the constructor for the class. I talk more about this in a moment.

Working Out the Cluster Rule of Thumb

Earlier, I established that you could quickly estimate the optimum number of clusters to generate. I've included a method to give you the number of clusters to generate based on the instance rows.

```
public int calculateRuleOfThumb(int rows) {
    return (int)Math.sqrt(rows/2);
}
```

The number of rows is passed in as an integer variable. You return the square root of the row count divided by two. If you already know how many clusters you want, just hard-code that number.

Building the Cluster

The `main` constructor handles the building of the cluster using the Weka API. As in the previous examples, you're using the `SimpleKMeans` class to build the cluster.

It's a small block of code. Weka handles things for you, so there's not a large amount of preparation to do.

```
public WekaCluster(String filepath) {
        try {
            Instances data = DataSource.read(filepath);
            int clusters = calculateRuleOfThumb(data.numInstances());
            System.out.println("Creating k-means model with "
+ clusters + " clusters.");
            SimpleKMeans kMeans = new SimpleKMeans();
            kMeans.setNumClusters(clusters);
            kMeans.buildClusterer(data);

            showCentroids(kMeans);
            showInstanceInCluster(kMeans, data);

            testRandomInstances(kMeans);
        } catch (Exception e) {
            e.printStackTrace();
        }

    }
```

The data is read in using the `DataSource.read()` method. This takes the string path name (or an InputStream is preferred) and saves the data as instances.

Next, you calculate the number of clusters to define using the method you created earlier with the rule of thumb calculation.

The actual building of the cluster is handled in the next four lines. The `SimpleKMeans` class is the same as the one used in the workbench and command line. You set the number of clusters you want to define (`setNumClusters()`) and a random number seed (`with setSeed()`) and then build the cluster.

Finally, you call two methods: one shows the location of the centroids of each cluster, and the second shows in which clusters the instances are located.

Printing the Centroids

Now that the model is built, you can start to show some results from it. First you print out the location of each cluster centroid.

```
public void showCentroids(SimpleKMeans kMeans) {
        Instances centroids = kMeans.getClusterCentroids();
        for(int i = 0; i < centroids.numInstances(); i++) {
            System.out.println("Centroid: " + i + ": " +
centroids.instance(i));
        }
    }
```

The `getClusterCentroids()` method returns a set of instances. It's a case of iterating through these and printing the result of each instance. As six clusters were created (via the rule of thumb method calculation), there should be six instances printed.

Printing the Cluster Information

To show which cluster the instance belongs to, the `showInstanceInCluster()` method takes the k-means model and the assigned instances. The code then iterates each of the instances and prints which it is assigned to based on the model.

```
    public void showInstanceInCluster(SimpleKMeans kMeans, Instances
  data) {
        try {
            for(int i = 0; i < data.numInstances(); i++) {
                System.out.println("Instance " + i + " is in cluster "
  + kMeans.clusterInstance(data.instance(i)));
            }
        } catch(Exception e) {
            e.printStackTrace();
        }
    }
```

Making Predictions

The program so far covers the creation of clusters and reporting the results of the instances. What happens when new data comes in? At present, you're not able to predict anything. It would be nice to have a method you can access that takes new values and predicts in which cluster the result would be grouped.

Instances can be created within code and then run against the clustering model to see where the new values would lie. You need to create another method to return the cluster prediction.

```
public int predictCluster(SimpleKMeans kMeans, double x, double y) {
        int clusterNumber = -1;
        try {
            double[] newdata = new double[] { x, y };
            Instance testInstance = new Instance(1.0, newdata);
            clusterNumber = kMeans.clusterInstance(testInstance);
        } catch (Exception e) {
            e.printStackTrace();
        }
        return clusterNumber;
    }
```

You're passing the model and the values of the x and y variables (in the same way the original data was in two attributes). A double array is created, and the two values are stored.

The `Instance` class is created. The first value is the weight that is to be assigned to the instance. This is a value between 0 and 1. The second value is the double array that you've just created with the x and y values.

In the same way that you showed the cluster by using the `clusterInstance()` method, you run the new instance and get the cluster number. This value is then returned to the calling method.

To test this, I'm going to create another method, which will iterate 100 times and generate random values. Obviously, in your code you'll be calling the predictor as required.

```
public void testRandomInstances(SimpleKMeans kMeans) {
        Random rand = new Random();
        for (int i = 0; i < 100; i++) {
            double x = rand.nextInt(200);
            double y = rand.nextInt(200);
            System.out.println(x + "/" + y +
" test in cluster " + predictCluster(kMeans, x, y));
        }
    }
```

The method is generating random numbers for the x and y values and passing them to the prediction method. Add this to the `main` constructor after the centroids and clusters are first printed by inserting the following line:

```
testRandomInstances(kMeans);
```

before the `catch` block is reached in the `WekaCluster` constructor.

The Final Code Listing

Here's the code assembled and ready to run:

```
import java.util.Random;

import weka.clusterers.SimpleKMeans;
import weka.core.Instance;
import weka.core.Instances;
import weka.core.converters.ConverterUtils.DataSource;
public class WekaCluster {

    public WekaCluster(String filepath) {
        try {
            Instances data = DataSource.read(filepath);

            int clusters = calculateRuleOfThumb(data.numInstances());
            System.out.println("Rule of Thumb Clusters = " + clusters);
```

```
            SimpleKMeans kMeans = new SimpleKMeans();
            kMeans.setNumClusters(clusters);
            kMeans.setSeed(42);
            kMeans.buildClusterer(data);

            showCentroids(kMeans, data);
            showInstanceInCluster(kMeans, data);

        } catch (Exception e) {
            e.printStackTrace();
        }
    }

    public int calculateRuleOfThumb(int rows) {
        return (int)Math.sqrt(rows/2);
    }

    public void showCentroids(SimpleKMeans kMeans, Instances data) {
        Instances centroids = kMeans.getClusterCentroids();
        for (int i = 0; i < centroids.numInstances(); i++) {
            System.out.println("Centroid: " + i + ": "
+ centroids.instance(i));
        }
    }

    public void showInstanceInCluster(SimpleKMeans kMeans, Instances
data) {
        try {
            for (int i = 0; i < data.numInstances(); i++) {
                System.out.println("Instance " + i + " is in cluster "
                        + kMeans.clusterInstance(data.instance(i)));
            }
        } catch (Exception e) {
            e.printStackTrace();
        }
    }

    public static void main(String[] args) {
        // Pass the arff location and the number of clusters we want
        WekaCluster wc = new
WekaCluster("/Users/Jason/kmeandata.arff");
    }

}
```

Running the Program

With the hard work done, you can run the program and inspect the results. From Eclipse, select Run ⇨ Run, and the program will start. The output in the console window should look something like this:

```
Rule of Thumb Clusters = 6
Centroid: 0: 11.8,65.9
Centroid: 1: 105.083333,118.333333
Centroid: 2: 68.933333,19.4
Centroid: 3: 81.6,106.6
Centroid: 4: 28.5,64
Centroid: 5: 43.913043,146.043478
Instance 0 is in cluster 0
Instance 1 is in cluster 0
Instance 2 is in cluster 0
Instance 3 is in cluster 0
Instance 4 is in cluster 0
Instance 5 is in cluster 0
Instance 6 is in cluster 0
Instance 7 is in cluster 0
Instance 8 is in cluster 0
Instance 9 is in cluster 0
Instance 10 is in cluster 4
....
```

As you can see, the rule of thumb calculation recommended creating six clusters. After executing the k-means clustering method, you displayed the centroid of each cluster in order to "eyeball" the distances of any data point in the cluster from its cluster's center; finally, you displayed cluster membership for each element of our original object data.

Finally, here's the output of the predictions and their output cluster class:

```
146.0/167.0 test in cluster 1
109.0/67.0 test in cluster 1
95.0/80.0 test in cluster 3
29.0/160.0 test in cluster 5
165.0/193.0 test in cluster 1
33.0/167.0 test in cluster 5
108.0/73.0 test in cluster 1
63.0/63.0 test in cluster 2
186.0/176.0 test in cluster 1
67.0/47.0 test in cluster 2
43.0/5.0 test in cluster 2
85.0/9.0 test in cluster 2
152.0/60.0 test in cluster 1
```

Further Development

You will discover putting together a basic cluster algorithm with SimpleKMeans is a fairly straightforward matter. I've covered the main aspects of coding a solution. There are obvious developments from this point, such as connecting to a database table with Java Database Connectivity (JDBC) and extracting the data into instances.

One thing to remember with Weka is that when huge volumes of data are applied, the memory performance can suffer. I suggest that most needs of enterprises are still covered using this method and can be developed with scale in mind. In particular, sampling the data to fit in Weka memory will give you very good results.

Summary

Clustering will be one of those machine learning techniques that you'll pull out again and again. To that end, it does need some thought before you go building clusters and seeing what happens.

You've created simple k-means clusters in Weka via the workbench, on the command line, and within code. Obviously, there are plenty of options from this point on, but with what you've read in this chapter, you'll be able to get a system up and working quickly.

Association Rules Learning

Among the machine learning methods available, association rules learning is probably the most used. From point-of-sale systems to web page usage mining, this method is employed frequently to examine transactions. It finds out the interesting connections among elements of the data and the sequence (behaviors) that led to some correlated result.

This chapter describes how association rules learning methods work and also goes through an example using Apache Mahout for mining baskets of purchases. This chapter also touches on the myth, the reality, and the legend of using this type of machine learning.

Where Is Association Rules Learning Used?

The retail industry is tripping over itself to give you, the customer, offers on merchandise it *thinks* you will buy. To do that, though, it needs to know what you've bought previously and what other customers, similar to you, have bought. Brands such as Tesco and Target thrive on basket analysis to see what you've purchased previously. If you think the amount of content that Twitter produces is big, then just think about point-of-sale data; it's another world. Some supermarkets fail to adopt this technology and never look into baskets, much to their competitive disadvantage. If you can analyze baskets and act on the results, then you can see how to increase bottom-line revenue.

Association rules learning isn't only for retail and supermarkets, though. In the field of web analytics, association rules learning is used to track, learn, and predict user behavior on websites.

There are huge amounts of biological data being mined to gain knowledge. Bioinformatics uses association rules learning for protein and gene sequencing. It's on a smaller scale compared to something like computational biology, as it homes in on specifics compared to something like DNA. So, studies on mutations of genomes are part of a branch of bioinformatics that's probably working with it.

Web Usage Mining

Knowing which pages a user is looking at and then suggesting which pages might be of interest to the user is commonplace to keep a website more compelling and "sticky." For this type of mining, you require a mechanism for knowing which user is looking at which pages; the user could be identified by a user session, a cookie ID, or a previous user login where sites require users to log in to see the information.

If you have access to your website log files, then there is opportunity for you to mine the information. Many companies use the likes of Google Analytics as it saves them mining logs themselves, but it's worthwhile doing your own analysis if you can.

The basic log file, for example, has information against which you could run some basic association rules learning. Looking at the Apache Common Log Format (CLF), you can see the IP address of the request and the file it was trying to access.

```
86.78.88.189 - thisuserid [10/May/2014:13:55:59 -0700]
"GET /myinterestingarticle.html HTTP/1.0" 200 2326
```

By extracting the URL and the IP address, the association rules could eventually suggest related content on your site that would be of interest to the user.

Beer and Diapers

It is written on parchment dating back many years, the parable of the beer and the diapers (or nappies, as I will always call them).

Tis written on this day that the American male of the species would frequent the larger markets of super the day prior to the Sabbath. Newly attired with sleeping eyes and new child, said American male would buy device of child's dropping catching of cloth and safety pin, when, lo, he spotteth the beer of delights full appreciating he shall not make it to the inn after evensong, such be his newly acquired fatherly role. And Mart of Wal did look upon this repeated behavior and move the aisles according to the scriptures of the product of placement, thus increasing the bottom line.

This story has been preached by marketing departments the world over (possibly not in the style presented here), and it's been used in everything from keynotes to short talks, from hackathons to late night code jams. However, it's a case of fact mixed with myth.

When he was CEO of a company called Mindmeld, Thomas Blischok was also on the panel of a webcast on the past, present, and future of data mining and had managed the study on data that spawned the beer and nappies story. The study went back to the early 1990s when his team was looking at the basket data for Osco Drug. They did see a correlation on basket purchases between 5 p.m. and 7 p.m. and presented the findings to their client.

After that point, there's some confusion about where the story actually goes. Many versions are basically myth and legend; they've generated great chat and debate around the water cooler for years and will continue to do so. I had the pleasure of meeting Mark Madsen from Third Nature who also has a large interest in the story; it's amazing how this story has been borrowed, used, referenced, and carried forward. Ultimately, it was just good use of Structured Query Language (SQL).

The myth has now been superseded by the privacy-fearing consumer story known as the "Target can predict whether I'm pregnant or not" scenario. I, for one, have two reasons why Target could never predict my outcome: I've never shopped there, and it's biologically impossible. You don't need a two-node decision tree to figure that out. (Read Chapter 5, "Working with Decision Trees," for more information on that subject.)

NOTE   For the full story on the beer and diapers legend, take a look at D.J. Power's article from November 2002 at

`http://www.dssresources.com/newsletters/66.php`

The myth will live on forever, I'm sure (especially if you're in marketing), and it makes for good reading.

How Association Rules Learning Works

The basket analysis scenario is a good example to explain with, so I'll continue with it. Consider the following table of transactions:

TRANSACTIONID	PRODUCT1	PRODUCT2	PRODUCT3	PRODUCT4
1	True	True	False	False
2	False	False	True	False
3	False	False	False	True

TRANSACTIONID	PRODUCT1	PRODUCT2	PRODUCT3	PRODUCT4
4	True	True	True	False
5	False	True	False	False

This is essentially an item set of transactions with a transaction ID and the products (could be milk, nappies, beer, and beans, for example).

Ultimately you're looking for associations in the products. For example, if a customer buys products 1 and 2, he is likely to buy product 4.

So here is a set of items:

$$I = \{product1, product2, product3, product4\}$$

And here is a set of transactions:

$$T = \{product1, product2, product4\}$$

Each transaction must have a unique ID for the rule to glean any information. Also, it's worth noting that this sort of rule needs hundreds of transactions before it starts to generate anything of value to you. The larger the transaction set, the better the statistical output will be and the better the predictions will be.

The rule is defined as an implication; what you're looking at is the following:

$$X, Y \subseteq I, \text{where} X \cap Y = (/).$$

In plain English, what you're saying is X and Y are a subset of the item set in the intersection of X and Y.

They take on the form of a set as items denoted as X and Y. In scary math books, it will look like this: $X \Rightarrow Y$. The X denotes the items set before (or left of) the rule, called the *antecedent*, and the Y is the item set after (or right of) the rule, called the *consequent*.

Getting back to the products in the basic item set:

$$I = \{product1, product2, product3, product4\}$$

The true/false statements show whether the item is in that basket transaction or not.

To get the true picture of how the rules work, you need to investigate a little further into the concepts of support, confidence, lift, and conviction.

Support

Support is defined as the proportion of items in the data that contain the item set. It's written like so:

$$\text{Supp}(X) = \frac{transactions\_containing\_X}{total\_number\_of\_transactions}$$

If you were to take transaction number 1, as an example, you'd have the following equation:

$$\text{supp}(X) = \frac{\{product1, product2\}}{5} = 0.2$$

The item set appears only once in the transaction log, and there are five transactions, so the support is 1/5, which is 0.2.

Confidence

Confidence in the rule is measured as

$$conf(X \Rightarrow Y) = \text{supp}(X \cap Y) / \text{supp}(X)$$

What you're defining here is the proportion of transactions containing set X that also contain Y. This can be interpreted as the probability of finding the right-hand side of transactions under the condition of finding them on the left-hand side.

To use an analogy, think of parimutuel betting. All bets are placed together in a pool, and after the race has finished, the payout is calculated based on the total pool (minus the commission to the agent). For example, assume there are five horses racing and bets have been placed against each one:

HORSE NUMBER	BET
1	$40
2	$150
3	$25
4	$40
5	$30

The total pool comes to $285, and after the event is run and the winner is confirmed, the payout can be calculated. Assuming that horse number 4 was the winner, the calculation would be as follows:

Pool size after commission = $285 × (1 − 0.15) = $242.25

Payout per $1 on outcome 4 = $6.05 per $1 wagered

Lift

Lift is defined as the ratio of the observed if the X and Y item sets were independent. It's written as follows:

$$Lift(X \Rightarrow Y) = \frac{\text{supp}(X \cup Y)}{\text{supp}(X) \times \text{supp}(Y)}$$

Conviction

Finally there's *conviction*, which is defined as the ratio of the expected frequency that X occurs without Y:

$$conv(X \Rightarrow Y) = \frac{1 - \text{supp}(Y)}{1 - conf(X \Rightarrow Y)}$$

Defining the Process

Association rules are defined to satisfy two user-defined criteria, a minimum support value and a minimum confidence. The rules generation is done in two parts.

First the minimum support is applied to all the frequent item sets in the database (or file or data source). The frequent item sets along with the minimum confidence are used to form the rules.

Finding frequent item sets can be hard; it involves trawling through all the possible item combinations in the item sets. The number of possible item sets is the "power set" over the item set.

For example, if you have the following:

$$I = \{p1, p2, p3\}$$

then the power set of I would be this:

$$\{\{p1\}, \{p2\}, \{p3\}, \{p1, p2\}, \{p1, p3\}, \{p2, p3\}, \{p1, p2, p3\}\}$$

Notice that the empty set ({}) is omitted in the power set; this formulation gives you a size of 2^n-1, where n is the number of items. A small increase in the number of items causes the size of the power set to increase enormously; therefore, this method is quite hungry in memory when using something like the Apriori algorithm. Obviously, the power set of all combinations of baskets does not occur, and the calculation will be based only on those basket combinations that do. Nonetheless, it is still very expensive in time and memory to run calculations based on this method.

Algorithms

There are several algorithms used in association rule learning that you'll come across; the two described in this section are the most prevalent.

Apriori

Using a bottom-up approach, the Apriori algorithm works through item sets one at a time. Candidate groups are tested against the data; when no extensions to the set are found, the algorithm will stop. The support threshold for the example is 3.

Consider the following item set:

{1,2,3,4}

{1,3,4}

{1,2}

{2,3,4}

{3,4}

{2,4}

First, it counts the support of each item:

{1} = 3

{2} = 5

{3} = 4

{4} = 5

The next step is to look at the pairs:

{1,2} = 2

{1,3} = 2

{1,4} = 2

{2,3} = 2

{2,4} = 3

{3,4} = 4

As {1,2}, {1,3}, {1,4}, and {2,3} are under the chosen support threshold you can reject them from the triples that are in the database. In the example, there is only one triple:

{2,3,4} = 1 (we've discounted one from the {1,2} group)

From that deduction, you have the frequent item sets.

FP-Growth

The Frequent Pattern Growth (FP-Growth) algorithm works as a tree structure (called an FP-Tree). It creates the tree by counting the occurrences of the items in the database and storing them in a header table.

In a second pass, the tree is built by inserting the instances it sees in the data as it goes along the header table. Items that don't meet the minimum support threshold are discarded; otherwise, they are listed in descending order.

You can think of the FP-Growth algorithm like a graph, as is covered in the chapter about Bayesian Networks (Chapter 4). With a reduced dataset in a tree formation, the FP-Growth algorithm starts at the bottom—the place with the longest branches—and finds all instances of the given condition. When no more single items match the attribute's support threshold, the growth ends, and then it works on the next part of the FP-Tree.

Mining the Baskets—A Walk-Through

There are a few libraries that deal with association rule learning; in this chapter, I'll be once again using Weka and moving away from Mahout, which I covered in the first edition of this book. With the Apache Spark MLLib project, there are the FPGrowth and association rule learning packages, but I will save those for the Spark, which is covered in Chapter 13.

The Raw Basket Data

Within the Weka application distribution there is a test training set of basket data, so to keep things simple, I'm going to use that. If you open the `supermarket.arff` file, you will see the standard Weka data representation. Each basket item is an attribute,

```
@attribute 'tea' { t}
@attribute 'biscuits' { t}
@attribute 'canned fish-meat' { t}
```

```
@attribute 'canned fruit' { t}
@attribute 'canned vegetables' { t}
@attribute 'breakfast food' { t}
@attribute 'cleaners-polishers' { t}
@attribute 'soft drinks' { t}
@attribute 'health food other' { t}
@attribute 'beverages hot' { t}
@attribute 'health&beauty misc' { t}
@attribute 'deodorants-soap' { t}
@attribute 'mens toiletries' { t}
@attribute 'medicines' { t}
@attribute 'haircare' { t}
@attribute 'total' { low, high} % low < 100
```

There are two attribute values, either true or false (which is represented as a space character). If the value is true, then the attribute has been purchased for that transaction. The last line of the attribute list is the basket value, whether it was high or low.

After the line with `@data`, you then have all the basket transactions. One line has all the attributes separated by commas. Where there is a `t`, that indicates a purchase.

```
?,?,?,?,?,?,?,?,?,?,?,t,t,t,?,t,?,t,?,?,t,?,?,?,t,t,t,t,?,t,?,t,t,?,?,?,?,
?,?,t,t,t,?,?,?,?,?,?,?,t,?,?,?,?,?,?,?,?,?,t,?,t,?,?,t,?,t,?,?,?,?,?,?,?,?,
?,?,?,?,?,?,?,t,?,?,t,t,?,?,?,?,?,?,?,?,?,?,?,?,?,?,?,?,?,?,?,?,?,?,?,?,?,?,
?,?,?,?,?,?,?,?,t,?,?,?,?,?,?,?,?,?,?,?,?,?,?,?,?,?,?,?,?,?,?,?,?,?,?,?,?,?,
?,?,?,?,?,?,?,?,?,?,?,?,?,?,?,?,?,?,?,?,?,?,?,?,?,?,?,?,?,?,t,?,?,?,?,?,?,?,
?,?,?,?,?,?,?,?,?,?,?,?,?,?,?,?,?,?,?,?,?,?,?,?,?,?,?,high
```

Using the Weka Application

Open the Weka application and from the menu select the Explorer options. This will take you to the main Weka workbench where all the work will be done (see Figure 7.1).

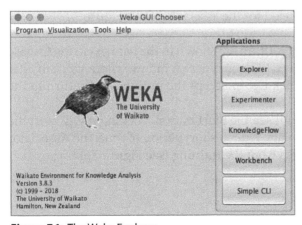

Figure 7.1: The Weka Explorer

The data file is in the data directory of the Weka distribution. Click the Open File button and locate the data file `supermarket.arff` (see Figure 7.2).

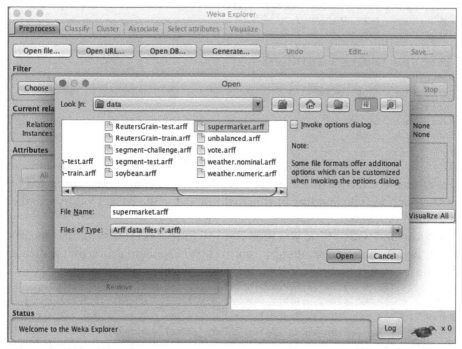

Figure 7.2: Weka File Explorer

Once the file is open, you will see the Preprocess pane appear. There should be 4,627 instances of basket data. In the attributes pane are the basket items, the departments, and the total. With the data looking right, it's time to click the Associate tab (see Figure 7.3).

Within the Associate pane the main work of the algorithm is done. From here you can change the options of the algorithm. If you click the Choose button, you can change the algorithm used for the association (see Figure 7.4). For now let's keep with Apriori.

The settings for the Apriori algorithm are shown in the bar next to the Choose button. If you click the bar, you will open a new window where you can alter the options (see Figure 7.5). Click OK to accept the options and return to the Associate pane.

To the left of the pane you will see the Start button; clicking it will start the learning process on the basket data. After a short period of time, the Associator output pane will report findings from the training (see Figure 7.6).

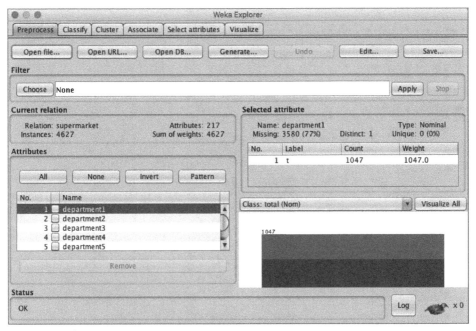

Figure 7.3: The Data Preprocess section

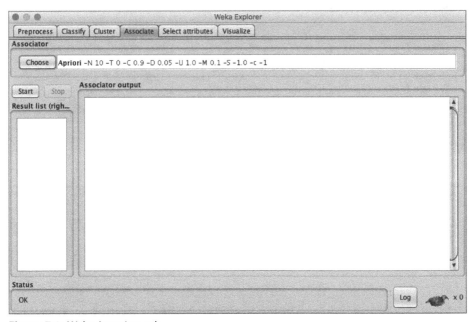

Figure 7.4: Weka Associate tab

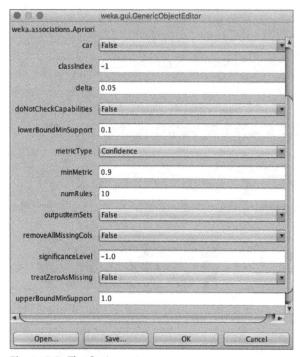

Figure 7.5: The Options pane

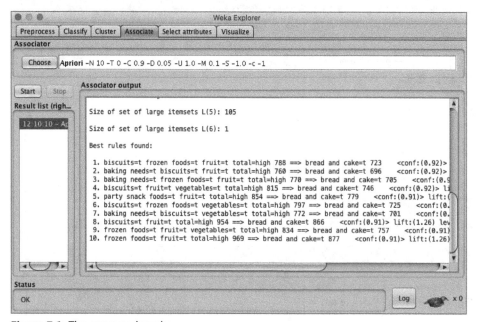

Figure 7.6: The generated results

Inspecting the Results

As we asked for 10 rules in the options and there is sufficient data to get a prediction, the output specifies the items that are frequently used. As you can see, "bread and cake" has a repeated high confidence and lift.

```
Size of set of large itemsets L(5): 105
Size of set of large itemsets L(6): 1
Best rules found:

 1. biscuits=t frozen foods=t fruit=t total=high 788 ==> bread and
cake=t 723    <conf:(0.92)> lift:(1.27) lev:(0.03) [155] conv:(3.35)
 2. baking needs=t biscuits=t fruit=t total=high 760 ==> bread and
cake=t 696    <conf:(0.92)> lift:(1.27) lev:(0.03) [149] conv:(3.28)
 3. baking needs=t frozen foods=t fruit=t total=high 770 ==> bread and
cake=t 705    <conf:(0.92)> lift:(1.27) lev:(0.03) [150] conv:(3.27)
 4. biscuits=t fruit=t vegetables=t total=high 815 ==> bread and cake=t
746    <conf:(0.92)> lift:(1.27) lev:(0.03) [159] conv:(3.26)
 5. party snack foods=t fruit=t total=high 854 ==> bread and cake=t 779
<conf:(0.91)> lift:(1.27) lev:(0.04) [164] conv:(3.15)
 6. biscuits=t frozen foods=t vegetables=t total=high 797 ==> bread and
cake=t 725    <conf:(0.91)> lift:(1.26) lev:(0.03) [151] conv:(3.06)
 7. baking needs=t biscuits=t vegetables=t total=high 772 ==> bread and
cake=t 701    <conf:(0.91)> lift:(1.26) lev:(0.03) [145] conv:(3.01)
 8. biscuits=t fruit=t total=high 954 ==> bread and cake=t 866
<conf:(0.91)> lift:(1.26) lev:(0.04) [179] conv:(3)
 9. frozen foods=t fruit=t vegetables=t total=high 834 ==> bread and
cake=t 757    <conf:(0.91)> lift:(1.26) lev:(0.03) [156] conv:(3)
10. frozen foods=t fruit=t total=high 969 ==> bread and cake=t 877
<conf:(0.91)> lift:(1.26) lev:(0.04) [179] conv:(2.92)
```

Let's break these rules down a little.

```
1. biscuits=t frozen foods=t fruit=t total=high 788 ==> bread and cake=t
723    <conf:(0.92)> lift:(1.27) lev:(0.03) [155] conv:(3.35)
```

We have a format that's presented, and the => tells us that there are four items that were preceding (were the antecedent of) the consequent value of "bread and cake." Biscuits, frozen food, fruit, and a high total value were reported 788 times in the dataset. The result also appeared 723 times.

There's a confidence value of 92 percent (0.92 in the output). The minMetric setting in the algorithm options means we can tune this value; the default is set to 0.9. This is useful when working with large sets of data while doing discovery phases of a project. Working with a lower value and working up until the results start to look refined and useful to the project.

Summary

Association rules learning is domain specific, so how it will work in your organization will vary from case to case. This chapter offered a brief overview and a working demo with Weka. Other libraries are available, so it's also worth taking the time to evaluate them and see if they fit with your strategy.

# CHAPTER 8

Support Vector Machines

With most machine learning tasks, the aim is usually to classify something into a group that you can then inspect later. When it's a couple of class types that you're trying to classify, then it's a trivial matter to perform the classification. When you are dealing with many types of classes, the process becomes more of a challenge. Support vector machines help you work through the challenging classifications.

This chapter looks at support vector machines: how the basic algorithm works in a binary classification sense, and then an expanded discussion on the tool.

What Is a Support Vector Machine?

A *support vector machine* is essentially a technique for classifying objects. It's a supervised learning method, so the usual route for getting a support vector machine set up would be to have some training data and some data to test the algorithm. With support vector machines, you have the linear classification—it's either that object or it's that object—or nonlinear. This chapter looks at both types.

There is a lot of comparison of using a support vector machine versus the artificial neural network, especially as some methods of finding minimum errors and the Sigmoid function are used in both.

It's easy to imagine a support vector machine as either a two- or three-dimensional plot with each object located within. Essentially, every object is a point in that space. If there's sufficient distance in the area, then the process of classifying is easy enough.

Where Are Support Vector Machines Used?

Support vector machines are used in a variety of classification scenarios, such as image recognition and handwriting pattern recognition.

Image classification can be greatly improved with the use of support vector machines. Being able to classify thousands or millions of images is becoming more and more important with the use of smartphones and applications like Instagram. Support vector machines can also do text classification on normal text or web documents, for instance.

Medical science has long used support vector machines for protein classification. The National Institute of Health has even developed a support vector machine protein software library. It's a web-based tool that classifies a protein into its functional family.

Some people criticize the support vector machine because it can be difficult to understand, unless you are blessed with a good mathematician who can guide and explain to you what is going on. In some cases you are left with a black-box implementation of a support vector machine that is taking in input data and producing output data, but you have little knowledge in between.

Machine learning with support vector machines takes the concept of a perceptron (as explained in Chapter 9) a little bit further to maximize the geometric margin. It's one of the reasons why support vector machines and artificial neural networks are frequently compared in function and performance.

The Basic Classification Principles

For those who've not immersed themselves in the way classification works, this section offers an abridged version. The next section covers how the support vector machine works in terms of the classification. I'm keeping the math as simple as possible.

Binary and Multiclass Classification

Consider a basic classification problem. You want to figure out which objects are squares and which are circles. These squares and circles could represent anything you want—cats and dogs, humans and aliens, or something else. Figure 8.1 illustrates the two sets of objects.

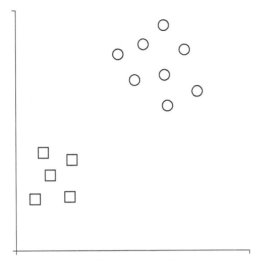

Figure 8.1: Two objects to classify

This task would be considered a binary classification problem, because there are only two outcomes; it's either one object or the other. Think of it as a 0 or a 1. With some supervised learning, you could figure out pretty quickly where those classes would lie with a reasonable amount of confidence.

What about when there are more than two classes? For example, you can add triangles to the mix, as shown in Figure 8.2.

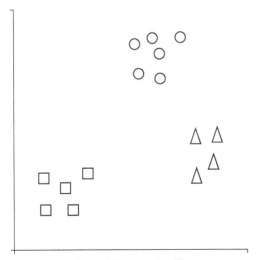

Figure 8.2: Three objects to classify

Binary classification isn't going to work here. You're now presented with a multiclass classification problem. Because there are more than two classes, you have to use an algorithm that can classify these classes accordingly. It's worth noting, though, that some multiclass methods use pair-wise combinations of binary classifiers to get to a prediction.

Linear Classifiers

To determine in which group an object belongs, you use a linear classifier to establish the locations of the objects and see if there's a neat dividing line—called a *hyperplane*—in place; there should be a group of objects clearly on one side of the line and another group of objects just as clearly on the opposite side. (That's the theory, anyway. Life is rarely like that, which is something that's covered more later in the chapter.) Assume that all your ducks are in a row. . .well, two separate groups.

As shown in Figure 8.3, visually it looks straightforward, but you need to compute it mathematically. Every object that you classify is called a point, and every point has a set of features.

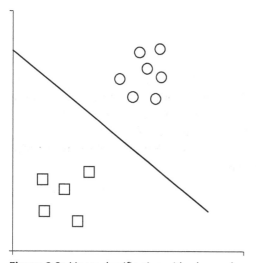

Figure 8.3: Linear classification with a hyperplane

For each point in the graph, you know there is an x-axis value and there is a y-axis value. The classification point is calculated as follows:

$$sign(ax + by + c)$$

The values for a, b, and c are the values that define the line; these values are ones that you choose, and you'll need to tweak them along the way until you get a good fit (clear separation). What you are interested in, though, is the result; you want a function that returns +1 if the result of the function is positive, signifying the point is in one category, and returns -1 when the point is in the other category. The function's resulting value (+1 or -1) must be correct for every point that you're trying to classify.

Don't forget that you have a training file with the correctly classified data so that you can judge the function's correctness; this approach is a supervised method of learning. This step has to be done to figure out where the line fits. Points that are further away from the line show more confidence that they belong to a specific class.

Confidence

You've just established that each point has a confidence based on its distance from the hyperplane line. The confidence can be translated into a probability. That gives the following equation:

$$P[l = +1 | x] = \frac{1}{1 + \exp(-(ax + by + c))}$$

This is for one point. What you need is the probability for every set of lines; these are then assigned to each of the objects in the training data.

$$\prod_{i=1}^{N} P[l_i | x_i] = \prod_{i=1}^{N} \frac{1}{1 + \exp(-l_i(ax_i + by_i + c))}$$

Probabilities are multiplied because the points have been drawn independently. You have an equation for each point that indicates how probable it is that a hyperplane is producing the correct categorization. Combining the probabilities for each point produces what is commonly defined as the "likelihood of the data"; you are looking for a number as close to 1 as possible.

Remember that probability is based on a value between 0 and 1. Within a set of objects, you're looking for a set of line parameters with the highest probability that confirms the categorization is correct.

Maximizing and Minimizing to Find the Line

Using a log function that is always increasing maximizes values that are above the equation. So, you end up with a function written as follows:

$$\sum_{i=1}^{N} -\log\left(1 + \exp\left(-l_i(ax_i + by_i + c)\right)\right)$$

To achieve minimization, you just multiply the equation by -1. It then becomes a "cost" or "loss" function. The goal is to find line parameters that minimize this function.

Linear classifiers are usually fast; they will process even large sets of objects with ease. This is a good thing when using them for document classification where the word frequencies might require measuring.

How Support Vector Machines Approach Classification

The basic explanation of linear classification is that the hyperplane creates the line that classifies one object and another. Support vector machines take that a step further.

Within the short space available, I outline how support vector machines work in both linear and nonlinear form. I also show you how to use Weka to do some practical work for you.

Using Linear Classification

Look at the set of circle and square objects again. You know how a hyperplane divides the objects into either 1 or -1 on the plane.

Extending that notion further, support vector machines define the maximum margin, assuming that the hyperplane is separated in a linear fashion. You can see this in Figure 8.4 with the main hyperplane line giving the following written notation:

$$w \bullet x - b = 0$$

This dot product shows the normal vector, and x is the point of the object. There is an offset of the hyperplane that goes from the origin to the normal vector.

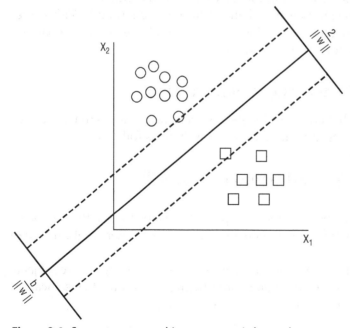

Figure 8.4: Support vector machines max margin hyperplane

As the objects are linearly separable, you can create another two hyper-planes—edge hyperplanes—that define the offset on either side of the main hyperplane. There are no objects within the region that spans between the main hyperplane and the edge hyperplanes.

On one side, there's the equation

$$w \bullet x - b = 1$$

and on the other side there's

$$w \bullet x - b = -1v$$

The objects that lie on the edge hyperplanes are the support vectors (see Figure 8.5).

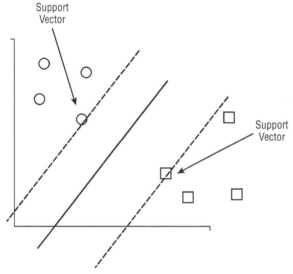

Figure 8.5: The support vectors on the hyperplane edges

When new objects are added to the classification, then the hyperplane and its edges might move. The key objective is to ensure a maximum margin between the +1 edge hyperplane and the -1 edge hyperplane.

If you can manage to keep a big gap between the categories, then there's an increase in confidence in your predictions. Knowing the values of the hyper-plane edges gives you a feel for how well your categories are separated.

After minimizing the value w (called $||w||$ in mathematical notation), you can look at optimizing w by applying the following equation:

$$\frac{1}{2} ||w||_2$$

Basically, you're taking half of ||w|| squared instead of using the square root of ||w||. Based on Lagrange multipliers, to find the maxima and minima in the function, you can now look for a saddle point and discount other points that don't match zero (fit inside the saddle).

> **NOTE**   For those that don't know, a saddle point is a mathematical function where you have two variables that meet at a critical point when both function values are zero. It's called a saddle point as that's the shape it produces in graphic form. You can read more about it at `http://wikipedia.org/wiki/Saddle_point`.

You're shaping the graph into a multidimensional space and seeing where the vectors lie in order to make the category distinctions as big as possible. With standard quadratic programming, you then apply the function expressing the training vectors as a linear combination

$$w = \sum_{i=1}^{n} \alpha_i y_i x_i.$$

Where αi is greater than zero, the xi value is a support vector.

Using Non-Linear Classification

In an ideal world, the objects would lie on one side of the hyperplane or the other. Life, unfortunately, is rarely like that. Instead, you see objects straying from the hyperplane, as shown in Figure 8.6.

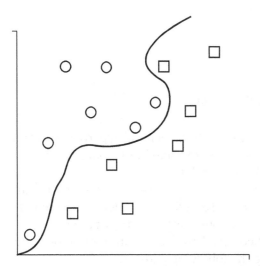

Figure 8.6: Objects rarely go where you want them to go.

By applying the kernel function (sometimes referred to as the "kernel trick"), you can apply an algorithm to fit the hyperplane's maximum margin in a feature space. The method is similar to the dot products discussed in the linear methods, but this replaces the dot product with a kernel function.

With a radial basis function, you have a few kernel types to choose from: the hyperbolic tangent, Gaussian radial basis function (or RBF, which is supported in Weka), and two polynomial functions—one homogenous and the other inhomogeneous.

The full scope of nonlinear classification is beyond the means of the introductory nature of this book. If you want to try implementing them, then look at the radial basis functions in the LibSVM classes when you use Weka. Now take a look at what Weka can do for you to perform support vector machine classification.

Using Support Vector Machines in Weka

Weka can classify objects using the support vector machines algorithm, but the implementation isn't complete and requires a download before you can use it. This section shows you how to set it up and run the support vector machines algorithm on some test data.

Installing LibSVM

The LibSVM library is an implementation of the support vector machines algorithm. It was written by Chih-Chung Chang and Chih-Jen Lin from the National Taiwan University. The library supports a variety of languages as well as Java including C, Python, .NET, MatLab, and R.

Weka LibSVM Installation

Using the LibSVM libraries within the Weka GUI application requires an installation of a JAR file. You can install LibSVM from GitHub. You can clone the binary distribution by running the following command (assuming you have Git installed):

```
git clone https://github.com/cjlin1/libsvm.git
```

The required files download into a clean directory.

You need to copy the `libsvm.jar` file to the same directory as your Weka installation directory (usually in the `/Applications` directory). You can easily drag and drop the file if desired; I work from the command line most of the time:

```
cp ./libsvm-3.18/java/libsvm.jar /Applications/weka-3-6-10
```

With the library in place, you can start Weka. If you are a Windows user just start Weka as normal, but if you run macOS or Linux, then you have to do it from the command line:

```
java -cp weka.jar:libsvm.jar weka.gui.GUIChooser
```

If you do not start Weka from the command line, then the classifier gives you an error to let you know that the SVM libraries were not found in the classpath.

With later versions of Weka, it's possible to install the LibSVM library from the package manager. The package manager is located from the main GUI Chooser application. From the list of packages, select the LibSVM package and click Install.

For using LibSVM in an application, the option is to either reference the JAR file in your classpath (which is covered further in this chapter) or use Maven and have the LibSVM library as a dependency reference.

```xml
<?xml version="1.0" encoding="UTF-8"?>
<project xmlns="http://maven.apache.org/POM/4.0.0"
         xmlns:xsi="http://www.w3.org/2001/XMLSchema-instance"
         xsi:schemaLocation="http://maven.apache.org/POM/4.0.0
http://maven.apache.org/xsd/maven-4.0.0.xsd">
    <modelVersion>4.0.0</modelVersion>

    <groupId>mlbook</groupId>
    <artifactId>Chapter8</artifactId>
    <version>1.0-SNAPSHOT</version>
    <dependencies>
        <!-- https://mvnrepository.com/artifact/nz.ac.waikato.cms.weka/
weka-stable -->
        <dependency>
            <groupId>nz.ac.waikato.cms.weka</groupId>
            <artifactId>weka-stable</artifactId>
            <version>3.6.7</version>
        </dependency>
        <!-- https://mvnrepository.com/artifact/tw.edu.ntu.csie/libsvm
-->
        <dependency>
            <groupId>tw.edu.ntu.csie</groupId>
            <artifactId>libsvm</artifactId>
            <version>3.23</version>
        </dependency>
    </dependencies>
</project>
```

A Classification Walk-Through

You will see the GUI Chooser application open as you would when you open Weka by starting the GUI instead of using the command line (see Figure 8.7). Choose the Explorer option.

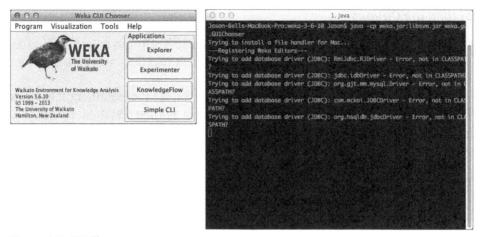

Figure 8.7: GUI Chooser

I'm going to use the 100,000 rows of vehicle data that is in the code and data repository that accompanies the book. Find the `.csv` file in the `data/ch08` folder and open it in Weka, as shown in Figure 8.8. Don't forget to change the file type from `.arff` to `.csv`.

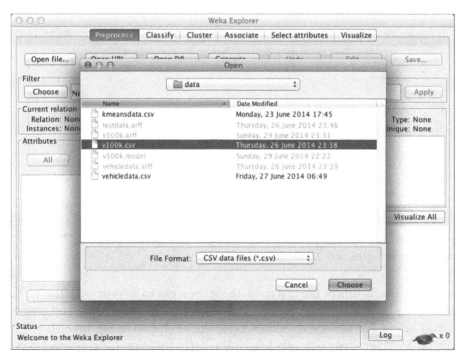

Figure 8.8: Loading the `.csv` file

Setting the Options

Click the Classify tab and then click the Choose button to select a different classification algorithm. Within the tree of algorithms, click Functions and then select LibSVM, as shown in Figure 8.9.

There are a couple of changes to make before you set the classifier off to work. First, you want a percentage split of training data against test data. In this case, you can be fairly confident that the data is not going to be difficult to classify and it's not going to be a nonlinear classification problem; you can train with 10 percent of the data (10,000 rows) and test with the 90 percent to see how it performs.

Click the Percentage Split option and change the default value of 66 percent to 10 percent, as shown in Figure 8.10.

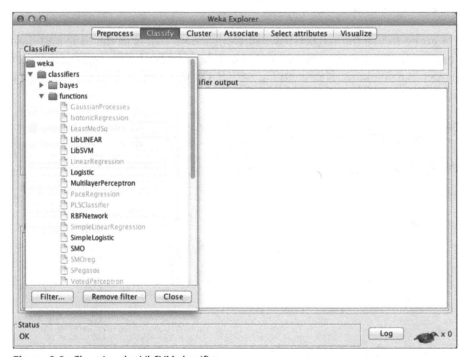

Figure 8.9: Choosing the LibSVM classifier

You want the results of the test data, the 90 percent to be output to the Weka console so you can see how it's performing. Click the Options button and ensure that the Output Predictions checkbox is ticked, as shown in Figure 8.11.

The LibSVM wrapper defaults to a radial basis function for its kernel type. Change that to the linear version you've been concentrating on by clicking the line with all the LibSVM options. This is located next to the Choose button within the Classifier pane.

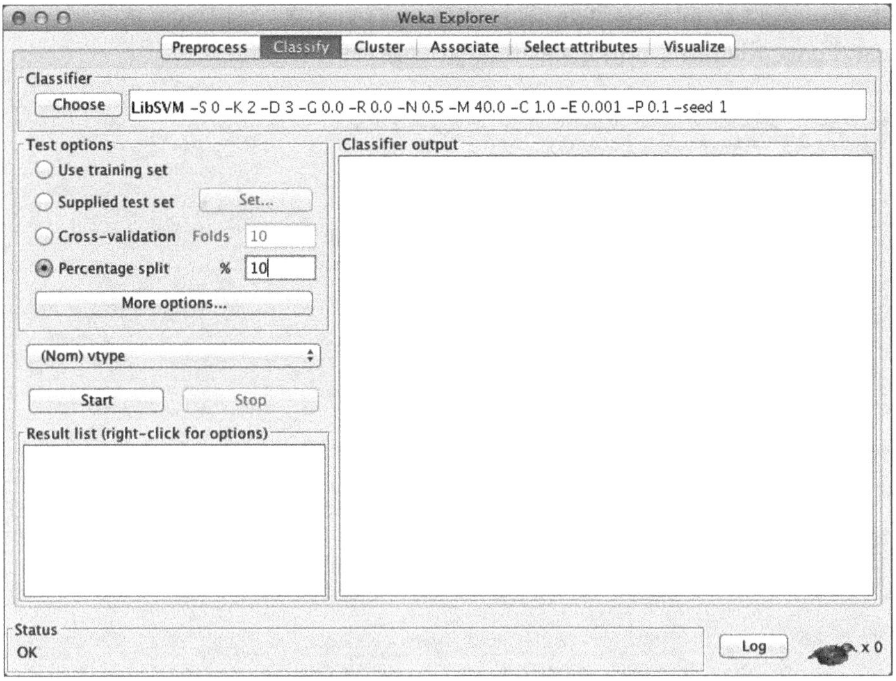

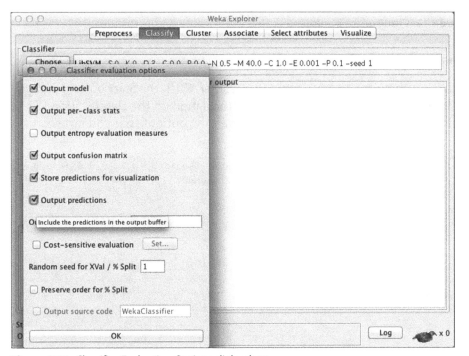

Figure 8.10: Changing the percentage split

Figure 8.11: Classifier Evaluation Options dialog box

Change the kernelType drop-down menu from Radial Basis Function to Linear. Leave the other options as they are (see Figure 8.12).

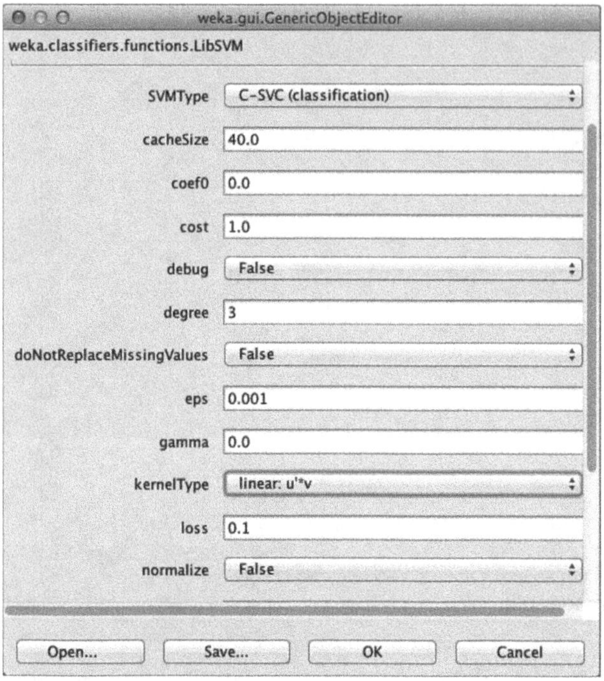

Figure 8.12: Changing the kernel type

Running the Classifier

With everything set, you can run the classifier. Click the Start button, and you see the output window start to output information on the classification.

First, you have the run information, or all the options that you just set.

```
=== Run information ===
Scheme:weka.classifiers.functions.LibSVM -S 0 -K 0 -D 3 -G 0.0 -R 0.0
  -N 0.5 -M 40.0 -C 1.0 -E 0.001 -P 0.1 -seed 1
Relation:     v100k
Instances:    100000
Attributes:   4
              wheels
              chassis
              pax
              vtype
Test mode:split 10.0% train, remainder test
```

Next, you get some general information on the classifier model.

```
=== Classifier model (full training set) ===
LibSVM wrapper, original code by Yasser EL-Manzalawy (= WLSVM)
Time taken to build model: 3.08 seconds
```

On my machine, the classifier trained on 10,000 instances in just over three seconds, which is 3,246 rows per second.

As you've set the output for the predictions to be shown, you get that next.

```
=== Predictions on test split ===
inst#,    actual, predicted, error, probability distribution
      1    4:Bike    4:Bike        0      0      0    *1
      2    4:Bike    4:Bike        0      0      0    *1
      3    3:Truck   3:Truck       0      0    *1      0
      4    1:Bus     1:Bus       *1      0      0      0
      5    1:Bus     1:Bus       *1      0      0      0
      6    2:Car     2:Car         0    *1      0      0
      7    4:Bike    4:Bike        0      0      0    *1
      8    3:Truck   3:Truck       0      0    *1      0
      9    3:Truck   3:Truck       0      0    *1      0
     10    2:Car     2:Car         0    *1      0      0
     11    4:Bike    4:Bike        0      0      0    *1
     12    2:Car     2:Car         0    *1      0      0
     13    3:Truck   3:Truck       0      0    *1      0
     14    3:Truck   3:Truck       0      0    *1      0
     15    4:Bike    4:Bike        0      0      0    *1
     16    1:Bus     1:Bus       *1      0      0      0
     17    1:Bus     1:Bus       *1      0      0      0
     18    2:Car     2:Car         0    *1      0      0
     19    1:Bus     1:Bus       *1      0      0      0
```

Based on the training data of 10,000, you've instructed Weka to try to predict the remaining 90,000 rows of data. The output window will have all 90,000 rows there, but the main things to watch out for are the actual and predicted results.

You get the evaluation on the test data showing the correct and incorrect assignments:

```
=== Evaluation on test split ===
=== Summary ===
Correctly Classified Instances       90000              100    %
Incorrectly Classified Instances         0                0    %
Kappa statistic                          1
Mean absolute error                      0
Root mean squared error                  0
Relative absolute error                  0    %
Root relative squared error              0    %
Total Number of Instances            90000
```

The confusion matrix shows the breakdown of the test data and how it was classified:

```
=== Confusion Matrix ===
    a     b     c     d   <-- classified as
22486     0     0     0 |   a = Bus
    0 22502     0     0 |   b = Car
    0     0 22604     0 |   c = Truck
    0     0     0 22408 |   d = Bike
```

Dealing with Errors from LibSVM

There are variations of the LibSVM library around the Internet and also different ways the random number generator handles numbers on differing operating systems. If you come across an error like the following:

```
java.lang.NoSuchFieldException: rand
java.lang.Class.getField(Unknown Source)
weka.classifiers.functions.LibSVM.buildClassifier(LibSVM.java:1618)
weka.gui.explorer.ClassifierPanel$16.run(ClassifierPanel.java:1432)
at java.lang.Class.getField(Unknown Source)
at weka.classifiers.functions.LibSVM.buildClassifier(LibSVM.java:1618)
at weka.gui.explorer.ClassifierPanel$16.run(ClassifierPanel.java:1432)
```

then it's worth looking at later versions of Weka with the new package manager (version 3.7 and later).

Saving the Model

You can save the model for this classification. On the result list, you see the date and time that the LibSVM classification was run. Right-click (Alt-click if you are a Mac user) functions.LibSVM and select Save Model. Find a safe place to save the model for future use.

Implementing LibSVM with Java

Using LibSVM within the Weka toolkit is easy to implement, but there comes a time when you'll want to use it within your own code so you can integrate it within your own systems.

Converting .csv Data to .arff Format

.csv files don't contain the data that Weka will need. You could implement the CSVLoader class, but I prefer to know that the .arff data is ready for use. It also makes it easier for others to decode the data model if they need to.

From the command line, you can convert the data from a .csv file to .arff in one command.

```
java -cp /Applications/weka-3-6-10/weka.jar \
weka.core.converters.CSVLoader v100k.csv > v100k.arff
```

To ensure that the conversion has worked, you can output the first 20 lines with the head command (your output should look like the following sample):

```
$ head -n 20 v100k.arff
@relation v100k

@attribute wheels numeric
@attribute chassis numeric
@attribute pax numeric
@attribute vtype {Bus,Car,Truck,Bike}

@data
6,20,39,Bus
8,23,11,Bus
5,3,1,Car
4,3,4,Car
5,3,1,Car
4,18,37,Bus
```

With everything looking fine, you can now set your attention on the Eclipse side of the project.

Setting Up the Project and Libraries

Using the same data, create a coded example with Java using Eclipse to create the project. Create a new Java Project (select File ➪ New ➪ Java Project) and call it MLLibSVM, as shown in Figure 8.13.

The Weka API and the LibSVM API need to be added to the project. Select File ➪ Properties and then select Java Build Path. Click the Add External JARs button. When the File dialog box displays, locate the weka.jar and libsvm.jar files and click Open (see Figure 8.14).

You have everything in place, so you can create a new Java class (File ➪ New ➪ Class) called MLLibSVMTest.java (see Figure 8.15) and put some code in place.

The basic code to get a support vector machine working in Weka is a fairly easy task.

Figure 8.13: Creating the new Java project

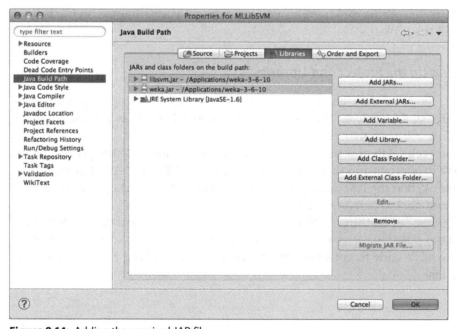

Figure 8.14: Adding the required JAR files

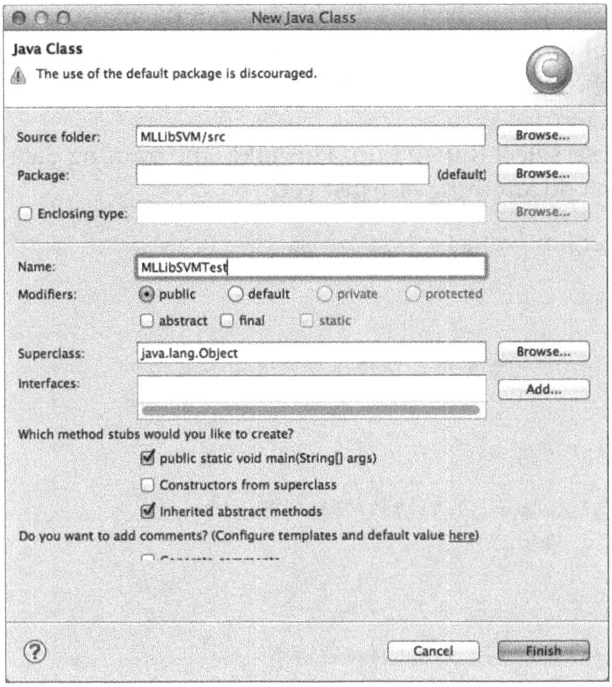

Figure 8.15: Creating a new Java class

```java
public class MLLibSVMTest {
    public MLLibSVMTest(String filepath){
        Instances data;
        try {
            data = DataSource.read(filepath);


            if (data.classIndex() == -1)
               data.setClassIndex(data.numAttributes() - 1);
            LibSVM svm = new LibSVM();
            String[] options = weka.core.Utils
             .splitOptions("-K 0 -D 3");
            svm.setOptions(options);
                svm.buildClassifier(data);
        } catch (Exception e) {
            e.printStackTrace();
        }
    }

    public static void main(String[] args) {
        MLLibSVMTest mllsvm =
         new MLLibSVMTest("/path/to/data/ch08/v100k.arff");
    }
}
```

There are a lot of option settings for the LibSVM library, but the main one I want to focus on is the kernel type. As in the Weka workbench, the default is the radial basis function. In the options, 2 designates this. For the linear kernel function, you change that to zero.

To run the code from Eclipse, select Run ➪ Run. This takes the training data and makes the model. It won't do anything else just yet.

```
Zero Weights processed. Default weights will be used
*
optimization finished, #iter = 9
nu = 7.999320068325541E-7
obj = -0.019999999949535163, rho = 2.1200468836658968
nSV = 4, nBSV = 0
*
optimization finished, #iter = 9
nu = 5.508757892156424E-7
obj = -0.013793103448275858, rho = -1.013793103448276
nSV = 5, nBSV = 0
*
optimization finished, #iter = 3
nu = 3.801428938130698E-7
obj = -0.009478672985781991, rho = 1.2180094786729856
nSV = 2, nBSV = 0
*
optimization finished, #iter = 5
nu = 1.8774340639289764E-7
obj = -0.004705882352941176, rho = -1.6070588235294119
nSV = 4, nBSV = 0
*
optimization finished, #iter = 6
nu = 8.90259889118131E-6
obj = -0.22222222222222227, rho = 1.6666666666666679
nSV = 3, nBSV = 0
*
optimization finished, #iter = 3
nu = 1.2308677001852457E-7
obj = -0.003076923076923077, rho = 1.1107692307692307
nSV = 2, nBSV = 0
Total nSV = 14
```

The output looks confusing, but what it is telling you is the number of support vectors (nSV), the number of bound support vectors (nBSV), and obj is the optimum objective value of the dual support vector machine.

Training and Predicting with the Existing Data

So far, you've trained with the full 100,000 lines of data from the .arff file. I want to train with 10 percent and then predict the remaining 90 percent in the same way as the workbench walkthrough.

The Weka API lets you add the options as you would in the workbench, so where you split the data for training, you can do the same within the code. Amend the options line and add the training split percentage like so:

```
String[] options = weka.core.Utils.splitOptions("-K 0 -D 3");
```

It now becomes the following:

```
String[] options = weka.core.Utils
    .splitOptions("-K 0 -D 3 -split-percentage 10");
```

To show the predictions of the data, add a new method that iterates through the instance data.

```
public void showInstanceClassifications(LibSVM svm, Instances data) {
        try {
            for (int i = 0; i < data.numInstances(); i++) {
                System.out.println("Instance " + i
                    + " is classified as a "
                        +
data.classAttribute().value((int)svm.classifyInstance(data.
                instance(i))));
            }
        } catch (Exception e) {
            e.printStackTrace();
        }
    }
```

The classifier always returns a numerical value as its result; it's up to you to turn that number into an integer and run it past the class attribute value to find out whether it's a bike, car, bus, or truck.

When you run the code again, you see the classifier generate as before with 10 percent of the training data, and then it classifies the whole data set.

```
Instance 99991 is classified as a Truck
Instance 99992 is classified as a Bus
Instance 99993 is classified as a Car
Instance 99994 is classified as a Truck
Instance 99995 is classified as a Car
Instance 99996 is classified as a Bus
Instance 99997 is classified as a Bike
Instance 99998 is classified as a Truck
Instance 99999 is classified as a Bike
```

Summary

This chapter was a whistle-stop tour of support vector machines. Whole books have been written on the subject, going deep into the intricacies of the vector machine and its kernel methods.

From a developer's point of view, treat this chapter as a launch pad for further investigation. In a practical scenario, you might gloss over the heavy theory and make Weka do the heavy lifting on a sample or subset of your data.

Artificial Neural Networks

There's something about gathering knowledge about the human brain that makes people tick. Many people think that if we can mimic how the brain works, we'll be able to make better decisions.

In this chapter, you look at how artificial neural networks work and how they are applied in the machine learning arena. If you are looking for how convolutional neural networks function, that will be covered in Chapter 11 when we look at image processing.

What Is a Neural Network?

Artificial neural networks are essentially modeled on the parallel architecture of animal brains, not necessarily human ones. The network is based on a simple form of inputs and outputs.

> ...a computing system made up of a number of simple, highly interconnected processing elements, which process information by their dynamic state response to external inputs.
>
> *Dr. Robert Hecht-Nielson as*
> *quoted in "Neural Network Primer:*
> *Part I" by Maureen Caudill, AI Expert, Feb. 1989*

In biology terms, a *neuron* is a cell that can transmit and process chemical or electrical signals. The neuron is connected with other neurons to create a network; picture the notion of graph theory with nodes and edges, and then you're picturing a neural network.

Within humans, there are a huge number of neurons interconnected with each other—tens of billions of interconnected structures. Every neuron has an input (called the *dendrite*), a cell body, and an output (called the *axon*), as shown in Figure 9.1.

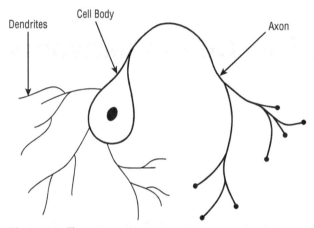

Figure 9.1: The neuron structure

Outputs connect to inputs of other neurons, and the network develops. Biologically, neurons can have 10,000 different inputs, but their complexity is much greater than the artificial ones I'm talking about here.

Neurons are activated when the electrochemical signal is sent through the axon. The cell body determines the weight of the signal, and if a threshold is passed, the firing continues through the output, along the dendrite.

Artificial Neural Network Uses

Artificial neural networks thrive on data volume and speed, so they are used within real-time or very near real-time scenarios. The following sections describe some typical use cases where artificial neural networks are used.

High-Frequency Trading

With the way artificial neural networks mimic the brain but with a much increased speed factor, they are perfect for high-frequency trading (HFT). Because HFT can make decisions far faster than a human can—thousands of transactions can be done in the same time it takes a human to make one—it's obvious why the majority of stock market systems have gone to the automated trading side.

High-frequency trading is usually done on a supervised learning method; there is a lot of training data available from which to learn. The artificial neural network is looking for entropy from the incoming data.

Credit Applications

Although many examples of credit applications are performed with decision trees, they are often run with artificial neural networks. With the variety of application data available, it's a fairly straightforward task to train the model to spot good and bad credit factors.

Data Center Management

Google uses neural networks for data center management. With incoming data on loads, operating temperatures, network equipment usage, and outside air temperatures, Google can calculate efficiency of the data center and be able to adjust the settings on monitoring and cooling equipment.

Jim Gao started this exercise as a Google 20 percent project (a program in which Google employees are encouraged to use 20 percent of their work time on their own projects) and, over time, has trained the model to be 99.6 percent accurate. If you are interested in reading more on this, check out Google's blog post on the subject here:

```
http://googleblog.blogspot.ca/2014/05/better-data-centers-through-
machine.html
```

Robotics

Artificial intelligence has been used in robotics for several years. Some artificial intelligence requires pattern recognition, and some requires huge amounts of sensor data to be fed into a neural network to determine what movement or action to take.

Training models in robotics takes an awful long time to create, mainly because there are potentially so many different inputs and output variables to process and learn from. For example, developers of autonomous driving vehicles need hundreds of thousands of hours of previous driving data to make a model that can handle many road conditions.

The car manufacturer Tesla collates the live driving data from its vehicles; this means it is generating knowledge 24 hours a day. This data is used to enrich the self-driving experience in its vehicles. Personally, I still prefer my hands on the wheel; that's just my preference.

Medical Monitoring

Medical machinery can be monitored via artificial neural networks, which involves the constant updating of many variables, such as heart rate, blood pressure, and so on. Conditions that have multiple variations and trigger symptoms can be calculated and monitored, and staff can be alerted when the variables go over certain thresholds.

There have been huge advances in medical imaging and using deep learning techniques such as convolutional neural networks to predict disease areas and support decision-making for the consultants and doctors.

Trusting the Black Box

There has been large-scale adoption of neural networks since the first edition of this book. While the volumes of data have increased, meaning that there is enough data to support accurate predictions, it's difficult to explain how these black-box algorithms are working.

The rule of thumb has been this: if you are unsure of the relationship between input and output of your model, then investigating with a neural network is a good way to go. If you have a good understanding of the input/output relationship, then chances are other traditional methods would be used.

I've always encouraged students, software professionals, and management to explore all the algorithmic options before settling on a neural network, especially if your findings will eventually end up scrutinized by another professional or the public at large.

Over time there has been much coverage on the incorrect predictions made by many strands of artificial intelligence. The term *explainable AI* has taken the forefront of the discussion when it comes to creating algorithms that are going to predict on behalf of others.

Furthermore, it's important you have sufficient volumes of training data for training neural networks. Small volumes of training will produce low accuracy scores, and there are times other machine learning algorithms, even linear regression, will perform better than a neural network.

It's also important to look at the quality of the training data. Does it evenly cover the spectrum of the question you are trying to answer? Biased data will give you incorrect predictions, and you will have no way of explaining why.

Ultimately, before you invest time, data, and money into a neural network, it's really worth asking yourself if this is the best way to do this task.

Breaking Down the Artificial Neural Network

Before you jump into data and a few examples, I will cover the rationale and working of the neural network.

Perceptrons

The basis for a neural network is the *perceptron*. Its role is quite simple. It receives an input signal and then passes the value through some form of function. It outputs the result of the function. (See Figure 9.2.)

Input   Output

2.5   1.5

Figure 9.2: A simple perceptron

Perceptrons deal with numbers when a number or vector of numbers is passed to the input. It is then passed to a function that calculates the outgoing value; this is called the *activation function*. The node can handle any number of inputs—Figure 9.3 shows two inputs passing into the function—and it takes the weighted sum of all the inputs.

Assuming the input is a vector Z, you'd end up with something like this:

Z 1 = 2

Z 2 = 5

Z 3 = 1

Or (2,5,1)

The weighted sum of all the inputs is calculated as follows:

$$\sum_i wiZi$$

In other words, "add it all up." So for the likes of me, who is not used to too much math notation, it looks like the following:

$$2_{w1} + 5_{w2} + 1_{w3}$$

The outgoing part of the node has a set threshold. If the summed value is over the threshold, then the output, denoted by the y variable, is 1, and if it's below the threshold, then y is 0 (zero).

You end up with the following equation:

$$if \sum_i wiZi \geq t\ then y = 1$$

$$else y = 0$$

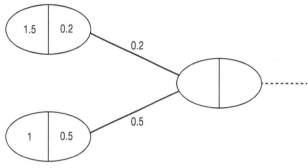

Figure 9.3: Perceptron with two inputs

The weight of the perceptron can be zero or any other value. If the weight value is zero, then it does not alter the input node value coming into the perceptron. Likewise, inputs can be positive or negative numbers. The key to the output is based on the weighted sum against the threshold.

That's the basis of a single-node perceptron. When you strip the components apart, it's quite basic in composition.

Activation Functions

The *activation function* is the processing that happens after the input is passed into the neuron. The result of this function determines whether the value is passed to the output axon and onto the next neuron in the network.

Commonly, the Sigmoid function (see Figure 9.4) and the hyperbolic tangent are used as activation functions to calculate the output.

The Sigmoid function outputs only one of two values: 0 and 1. For the programmers, the function is written as follows:

```
return 1.0 / (1.0 + Math.exp(-x));
```

The sharpness of the curve could also be altered if required, but for most applications a straight function is fine.

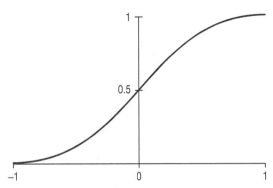

Figure 9.4: Sigmoid function

Multilayer Perceptrons

The problem with single-layer perceptrons is that they are linearly separable. The output is either one value or another.

If you think of an AND gate in logic theory, there is only one outcome if you have two inputs, as shown in Table 9.1.

Table 9.1: AND Gate Output Table

INPUT	OUTPUT
Off and On	Off
On and Off	Off
Off and Off	Off
On and On	On

The perceptron would be fashioned as shown in Figure 9.5.

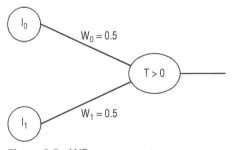

Figure 9.5: AND gate perceptron

The network output equation would be the following:

$$Output = \begin{cases} 1 \ if\left(W00 \times I0\right) + \left(W01 \times I1\right) > 0 \\ 0 \ Otherwise... \end{cases}$$

So far, I've covered the processing of one perceptron. Artificial neural networks have many interconnected neurons, each with its own input, output, and activation function.

For most machine learning functions, artificial neural networks are used for solving problems of a nonlinear fashion. Many problems cannot be solved in a purely linear fashion, so using a single-layer perceptron for this kind of problem-solving was never worth considering. If you think of an XOR gate (Exclusive OR) with the input types shown in Table 9.2, you could easily think of the network shown in Figure 9.6.

Table 9.2: Exclusive OR Output Table

INPUT	OUTPUT
Off and On	On
On and Off	On
Off and Off	Off
On and On	Off

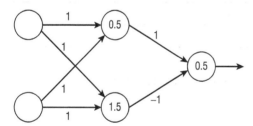

Figure 9.6: XOR gate network

Multilayer perceptrons have one or more layers between the input nodes and the eventual output nodes. The XOR example has a middle layer, called a *hidden layer*, between the input and the output (see Figure 9.7). Although you and I know what the outputs of an XOR gate would be (I've just outlined them in the table) and we could define the middle layer ourselves, a truly automated learning platform would take some time.

The question is, what happens in the hidden layer? Going back to the XOR example for a moment, you can see the two input nodes with their values. These would then be fed to the hidden layer, and the input is dependent on the output of the input layer.

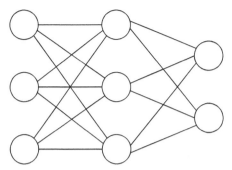

Figure 9.7: Multilayer perceptron with one hidden layer

This is where the neural network becomes useful. You can train the network for classification and pattern recognition, but it does require training. You can train an artificial neural network by unsupervised or supervised means.

The issue is that you don't know what the weight values should be for the hidden layer. By changing the bias in the Sigmoid function, you can vary the output layer, an error function can be applied, and the aim is to get the value of the error function to a minimum value.

I described the threshold function within the perceptron previously in the chapter, but this isn't suitable for your needs. You need something that is continuous and differentiable. With the bias option implemented in the Sigmoid function, each run of the network refines the output and the error function. This leads to a better-trained network and more reliable answers.

Back Propagation

Within the multilayer perceptron is the concept of *back propagation*, short for the "backward propagation of errors." Back propagation calculates the gradients and maps the correct inputs to the correct outputs.

There are two steps to back propagation: the propagation phase and the updating of the weight. This would occur for all the neurons in the network.

If you were to look at this as pseudocode—assuming an input layer, a single hidden layer, and an output layer—it would look like this:

```
initialize weights in network (random values)

while(examples to process)
  for each example x
    prediction = neural_output(network, x)
    actual = trained-output(x)
    error is (prediction - actual) on output nodes

backwardpass:
  compute weights from hidden layer to the output layer
  compute weights from input layer to hidden layer
```

```
update network weights
until all classified correctly against training data
return finalized network
```

Propagation happens, and the training is input through the network and generates the activations of the output. It then backward propagates the output activations and generates deltas of all the output and hidden layers of the network based on the target of the training pattern.

In the second phase, the weight update is calculated by multiplying the output delta and input activation. This gives you the gradient weight. The percentage ratio is then subtracted from the weight. The second part is done for all the weight axons in the network.

The percentage ratio is called the *learning rate*. The higher the ratio, the faster the learning. With a lower ratio you know the accuracy of the learning is good.

> **NOTE**   I appreciate that it's difficult to grasp mathematical concepts on neural networks in a book that focuses on the practical aspects of getting machine learning up and running quickly. This overview gives a general idea of how they work. The main concepts of input and output layers, perceptrons, and the notion of forward and backward propagation provide a good, although simple, grounding in the thought process.

Data Preparation for Artificial Neural Networks

For creating an artificial neural network, it's worth using a supervised learning method. However, this requires some thought about the data that you are going to use to train the network.

Artificial neural networks work only with numerical data values. So, if there are normalized things with text values, they need to be converted. This isn't so much an issue with the likes of gender, where the common output would be Male = 0 and Female = 1, for example. Raw text wouldn't be suitable, so it will either need to be tidied up, hashed to numeric values, or removed from the test data.

As with all data strategies, it's a case of thinking about what's important and what data you can live without.

As more variables increase in your data for classification, you will come across the phenomenon called "the curse of dimensionality." This is when added variables increase the total volume of training data required to get reasonable results and insight. So, when you are thinking of adding another variable, make sure you have enough training data to cover eventualities across all the other variables.

Although neural networks are pretty tolerant to noisy data, it's worth trying to ensure that there aren't large outliers that could potentially cause issues with the results. Either find and remove the wayward digits or turn them into missing values.

Artificial Neural Networks with Weka

The Weka framework supports a multilayer perceptron and trains it with the back propagation technique I just described. In this walk-through, you create some data and then generate a neural network.

Generating a Dataset

My dataset is going to contain classifications for different types of vehicles. I'm first going to create a Java program that generates some random, but weighted, data to give us four types of vehicles: bike, car, bus, and truck.

Here's the code listing:

```java
import java.io.BufferedWriter;
import java.io.FileWriter;
import java.io.IOException;
import java.util.Random;

public class MLPData {

    private String[] classtype = new String[] { "Bike", "Car", "Bus",
"Truck" };

    public MLPData() {

        Random rand = new Random(System.nanoTime());

        try {
            BufferedWriter out = new BufferedWriter(new FileWriter(
                    "vehicledata.csv"));
        out.write("wheels,chassis,pax,vtype\n");
            for (int i = 0; i < 100; i++) {
                StringBuilder sb = new StringBuilder();
                switch (rand.nextInt(3)) {
                case 0:
                    sb.append((rand.nextInt(1) + 1) + ",");
                    sb.append((rand.nextInt(1) + 1) + ",");
                    sb.append((rand.nextInt(1) + 1) + ",");
                    sb.append(classtype[0] + "\n");
                    break;
                case 1:
                    sb.append((rand.nextInt(2) + 4) + ",");
                    sb.append((rand.nextInt(4) + 1) + ",");
                    sb.append((rand.nextInt(4) + 1) +
                    sb.append(classtype[1] + "\n");
                    break;
                case 2:
                    sb.append((rand.nextInt(6) + 4) + ",");
                    sb.append((rand.nextInt(12) + 12) + ",");
```

```
sb.append((rand.nextInt(30) + 10) + ","); // passenger number

sb.append(classtype[2] + "\n");
                break;
        case 3:
            sb.append("18,"); // num of wheels
            sb.append((rand.nextInt(10) + 20) + ",");
            sb.append((rand.nextInt(2) + 1) +
            sb.append(classtype[3] + "\n");
            break;
        default:
            break;
        }
        out.write(sb.toString());

    }
    out.close();
} catch (IOException e) {
    e.printStackTrace();
}

}


public static void main(String[] args) {
    MLPData mlp = new MLPData();

}

}
```

When run, the preceding code creates a CSV file called `vehicledata.csv`. Start by creating 100 rows of output.

```
4,2,4,Car
9,20,25,Bus
5,14,18,Bus
5,2,1,Car
9,17,25,Bus
1,1,1,Bike
4,4,2,Car
9,15,36,Bus
1,1,1,Bike
5,1,4,Car
4,2,1,Car
```

As discussed previously, you need to perform a fair amount of training to make the neural network accurate in its predictions.

Loading the Data into Weka

Open the Weka toolkit and select the Explorer function to display the Explorer shown in Figure 9.8.

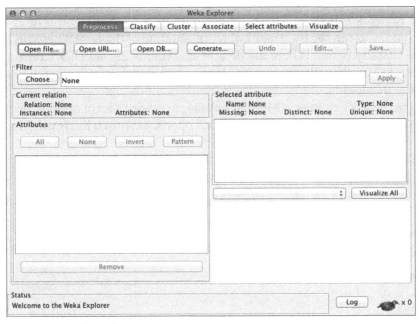

Figure 9.8: Weka Explorer

You're going to import the CSV file that's been created. Make sure that the Preprocess window is selected; then click the Open File button and select the `vehicledata.csv` file. Don't forget to change the File Format drop-down menu from `.arff` to `.csv`, as shown in Figure 9.9.

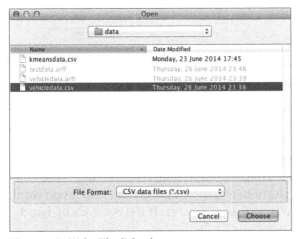

Figure 9.9: Weka File dialog box

You see the data loaded with the basic representation of the relation and attribute information.

Configuring the Multilayer Perceptron

The neural network function of Weka comes with its own graphic user interface. When run, you can see the graphical representation of the neural network.

Click the Classify panel. Where the default classifier is ZeroR, click Choose and change it to MultilayerPerceptron (see Figure 9.10), which is in the Functions branch of the tree listing.

You see the classifier change to MultilayerPerceptron with a lot of options next to it. If you click that line, a window of options opens, as shown in Figure 9.11.

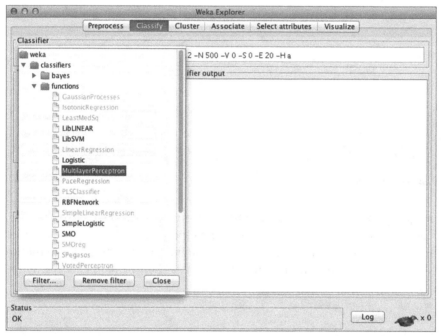

Figure 9.10: Changing the classifier

Change the GUI setting to True. This setting makes the neural network display in a graphic form; the display is also interactive, and you can change the network. If the GUI setting is set to False, then Weka generates the network for you without your intervention.

Although this version of the multilayer perceptron converts and handles your nominal values for you, it's still prudent to take the time to ensure that your data is prepared properly. The network autobuilds by default. If you want to create your own, then you can turn this off and craft the network by hand.

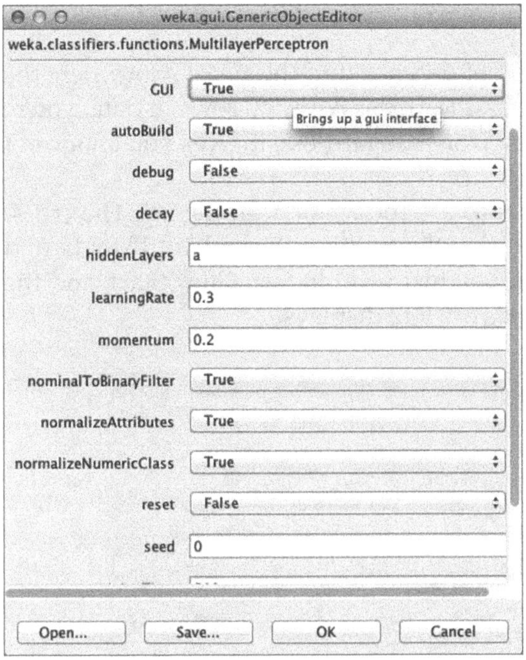

Figure 9.11: Options dialog box for MultilayerPerceptron

There are a few values that are worth keeping an eye on before you let the network do its training.

Learning Rate

The amount the weights are updated is defaulted at 0.3. If that seems a little heavy or too light, then you can adjust as desired.

Hidden Layers

You can define how many hidden layers the neural network will have. By default, Weka builds four (attributes and classes/2) (set to "a"), but you can also have just the attributes ("i"), just the classes ("o"), and the attributes and classes complete ("t").

Training Time

The number of epochs through which Weka iterates during training is set to 500. The higher the number, the lower the error rate will be. As you'll see in a moment, this can give varying results in the output.

When you are happy with the options, you can click OK and go back to the Classify window.

Training the Network

You have to do a few runs of neural networks to find the sweet spot where the network is coming up with good classifications. With 100 rows of data, you're not going to be solving much of any worth; regardless, it gives you an idea of how it works.

Make sure the test options are set to use the whole training set. The cross-validation is fine, but it ends up running the training through all 10 folds, and that can get time-consuming when you just want to test. Click Start, and the neural network window shown in Figure 9.12 displays.

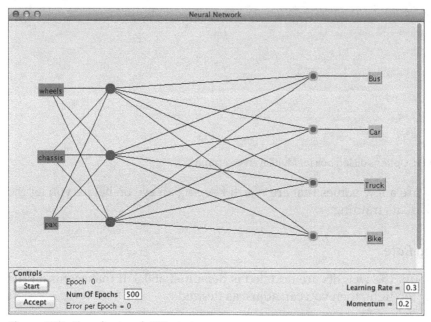

Figure 9.12: Neural network GUI window

Click Start, and you see the epoch count rise and the error rate decrease. If you click Accept by accident, then no data will have been classified, and the results will be wrong.

After the neural network has run, click the Accept button, and you will be returned to the classification output screen.

The full classifier output gives the output for the hidden layer nodes. Nodes 0, 1, 2, and 3, and the four nodes on the right side of Figure 9.12 are the output connections. The class attributes for classification are shown as bike, car, bus, or truck on the right side of the neural network output (refer to Figure 9.12).

```
Sigmoid Node 0
    Inputs      Weights
       Threshold     0.018993883149676594
```

```
        Node 4    -0.04038638643499096
        Node 5     0.0065483634965212145
        Node 6    -0.03873854654480489
Sigmoid Node 1
        Inputs    Weights
        Threshold  -0.0451840582741909
        Node 4    -0.002851224687941599
        Node 5    -0.012455737520358182
        Node 6    -0.0491382673800735
Sigmoid Node 2
        Inputs    Weights
        Threshold  -0.010479295335213488
        Node 4     0.02129170595398988
        Node 5     0.02877248387280648
        Node 6    -0.001813155428890656
Sigmoid Node 3
        Inputs    Weights
        Threshold   0.02680212410425596
        Node 4     0.006810392393573984
        Node 5    -0.04968676115705444
        Node 6    -0.015015642691489917
```

Nodes 4, 5, and 6 comprise the hidden layer that takes the input from the input attributes for wheels, chassis, and passenger count.

```
Sigmoid Node 4
        Inputs    Weights
        Threshold   0.011850776365702677
        Attrib wheels    0.0429940506718635
        Attrib chassis   -0.035625493582980464
        Attrib pax    -0.021284810000068835
Sigmoid Node 5
        Inputs    Weights
        Threshold   0.011165074786232076
        Attrib wheels    -0.018370069737576836
        Attrib chassis   -0.030938315802372954
        Attrib pax     0.01567513412449774
Sigmoid Node 6
        Inputs    Weights
        Threshold  -0.04753959806853169
        Attrib wheels    -0.00211881373779247
        Attrib chassis    0.040431974347463484
        Attrib pax    -0.017943250444400316
```

Each node has the input type and the weight values of the corresponding input node.

The summary shows how many instances have been correctly classified, along with other values for the error data if it has occurred.

In the last section, you can see how the classification counts added up in the Confusion Matrix, as shown here:

```
=== Confusion Matrix ===

   a  b  c  d    <-- classified as
  33  0  0  0 |   a = Bus
   0 27  0  0 |   b = Car
   0  0 20  0 |   c = Bike
   0  0  0 20 |   d = Truck
```

Altering the Network

With the GUI option set to True, you can add nodes and also remove input paths to parts of the hidden layer. If you make any changes, you need to retrain the neural network; the updated network will display in the GUI.

Which Bit Is Which?

Working from left to right on the GUI, you see the raw input nodes as labels in the yellow boxes. Red dots are the hidden layer nodes, and the orange dots are the output nodes. The orange labels are the classes with which the orange dot nodes are associated.

Adding Nodes

You can add a new node by clicking the GUI. The red dot appears to signify a hidden layer node. It won't be connected to anything, unless you have already selected nodes in the GUI.

Connecting Nodes

With the node selected, you can click another node to see the connection being made.

Removing Connections

To remove a connection, select one of the connected nodes and then right-click the other connected node. The connecting line disappears.

Removing Nodes

Right-clicking a node removes it and all the connections to it. Be careful to make sure that there aren't any other selected nodes; otherwise they, and their connections, will be removed, too.

Increasing the Test Data Size

Within the `for` loop of the `MLPData.java` program you created earlier in the chapter, change the loop count from 100 rows to 100,000 rows. Go back to the Preprocess window and load the new CSV file. It might take some time to load.

Now, go back to the Classify window and rerun the neural network. When the GUI window opens, you see the network looks the same as before in terms of the hidden layers. Where you had 500 epochs running against the 100 rows of data, you now have the same epoch number against all 100,000 rows of training data.

Click Start and the training begins. You'll notice a difference in response time from the GUI as it trains all 100,000 rows. The main thing to look at is the errors per epoch; the number keeps reducing to the point where you get minute changes per 100 to 200 epochs. By the time the training has finished, you will have a very accurate training model.

All this comes at a price of memory, though. My training set took more than two minutes.

```
Time taken to build model: 124.52 seconds
```

Two minutes isn't a huge amount of time in the grand scheme of things, but as I previously mentioned in regard to gathering data for neural networks, adding more variables gives the curse of dimensionality.

The more rows you can use for training, the better the prediction results will be. There is a point in time to figure out when there's too much training data against the errors per epoch. It takes some practice (and everyone's data is different, so there's no hard or fast rule), and it's a case of experiment, measure, and try again.

Implementing a Neural Network in Java

With the Weka API, you can build a neural network with the same multilayer perceptron that Weka uses within the GUI.

Creating the Project

Select File ⇨ New ⇨ Java Project and call it `MLPProcessor`, as shown in Figure 9.13.

You need to tell Eclipse where the Weka API is; it's called `weka.jar`. On macOS machines, Weka is usually installed within the Applications directory. The location on Windows machines varies depending on the specific operating system and Weka installation. In most cases, it will be `/Program Files (x86)/Weka-3-8/weka.jar`.

With the WekaCluster project selected, select File ⇨ Properties and look for the Java Build Path. Then click the Libraries tab. Add the external `jar` file by clicking Add External JARs; then in the file dialog box find the `weka.jar` file, as shown in Figure 9.14.

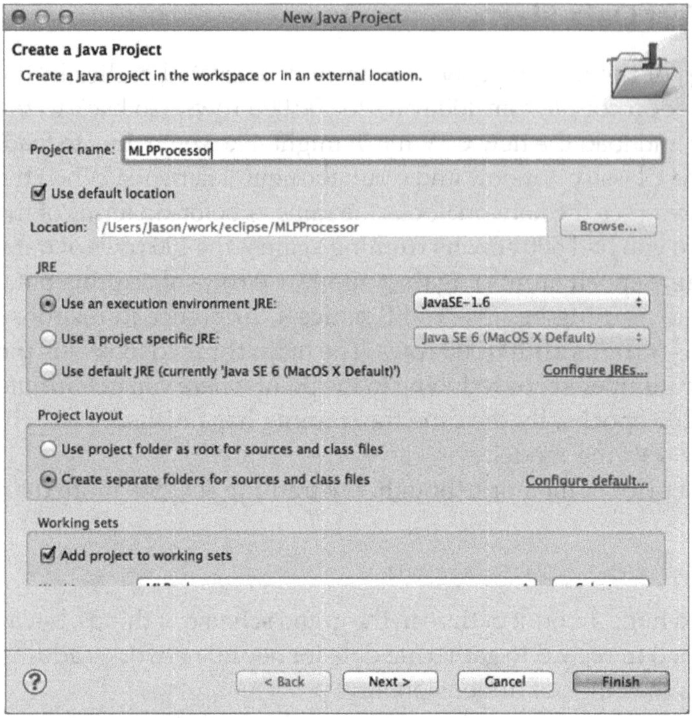

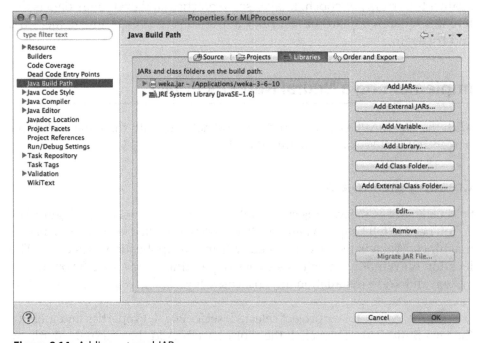

Figure 9.13: Eclipse New Project dialog box

Figure 9.14: Adding external JARs

The last thing to do is create a new class called `MLPProcessor.java` (using File ➭ New ➭ Class), as shown in Figure 9.15.

Figure 9.15: Creating a new class file

Writing the Code

The actual Java is straightforward. You're going to do the following:

1. Open the training data `.arff` file.
2. Create a multilayer perceptron and set the same options as the Weka GUI example.
3. Build the classifier.
4. Load some test data.
5. Run an evaluation test with the test data against the trained data.

You need to create a small test data file to test against the model. In a text file called `testdata.arff`, enter the following:

```
@relation vehicledata

@attribute wheels numeric
@attribute chassis numeric
```

```
@attribute pax numeric
@attribute vtype {Bus,Car,Truck,Bike}

@data
18,25,2,Truck
8,21,24,Bus
18,27,2,Truck
1,1,1,Bike
7,23,21,Bus
18,20,1,Truck
8,16,30,Bus
18,28,2,Truck
7,18,36,Bus
8,21,27,Bus
5,2,4,Car
18,28,1,Truck
5,1,1,Car
1,1,1,Bike
18,27,1,Truck
5,1,1,Car
6,15,38,Bus
7,21,38,Bus
18,20,2,Truck
1,1,1,Bike
18,28,2,Truck
18,24,2,Truck
18,20,1,Truck
1,1,1,Bike
5,17,18,Bus
18,27,1,Truck
4,4,3,Car
18,21,1,Truck
5,2,3,Car
4,3,3,Car
18,23,1,Truck
5,20,30,Bus
5,3,3,Car
18,28,1,Truck
5,3,1,Car
9,13,19,Bus
1,1,1,Bike
18,26,2,Truck
```

After you've created the test file, use the following code:

```
import java.io.FileNotFoundException;
import java.io.FileReader;
import java.io.IOException;
```

```java
import weka.classifiers.Evaluation;
import weka.classifiers.functions.MultilayerPerceptron;
import weka.core.Instances;
import weka.core.Utils;

public class MLPProcessor {

    public MLPProcessor() {
        try {
            FileReader fr = new FileReader("vehicledata.arff");

            Instances training = new Instances(fr);

            training.setClassIndex(training.numAttributes() -1);

            MultilayerPerceptron mlp = new MultilayerPerceptron();
            mlp.setOptions(Utils.splitOptions("-L 0.3 -M 0.2 -N 500
-V 0 -S 0 -E 20 -H 4"));


            mlp.buildClassifier(training);

            FileReader tr = new FileReader("testdata.arff");
            Instances testdata = new Instances(tr);
            testdata.setClassIndex(testdata.numAttributes() -1);

            Evaluation eval = new Evaluation(training);
            eval.evaluateModel(mlp, testdata);
System.out.println(eval.toSummaryString("\nResults\n*******\n", false));

            tr.close();
            fr.close();

        } catch (FileNotFoundException e) {
            e.printStackTrace();
        } catch (IOException e) {
            e.printStackTrace();
        } catch (Exception e) {
            e.printStackTrace();
        }
    }

    public static void main(String[] args) {
        MLPProcessor mlp = new MLPProcessor();
    }

}
```

The actual neural network is taken care of within three lines of code. Create the multilayer perceptron, set which class you want to determine, and then build the classifier. The rest of the code is loading the training and test data in.

Converting from CSV to Arff

CSV files don't contain the data that Weka needs. You could implement the CSVLoader class, but I prefer to know that the .arff data is ready for use. It also makes it easier for others to decode the data model if they need to.

From the command line, you can convert the data from a .csv file to .arff in one command.

```
java -cp /Applications/weka-3-6-10/weka.jar
weka.core.converters.CSVLoader \
vehicledata.csv > vehicledata.arff
```

If you inspect the .arff file, you see the attribute information set up for you.

```
@relation vehicledata

@attribute wheels numeric
@attribute chassis numeric
@attribute pax numeric
@attribute vtype {Bus,Car,Truck,Bike}

@data
6,20,39,Bus
8,23,11,Bus
5,3,1,Car
4,3,4,Car
5,3,1,Car
4,18,37,Bus
18,23,2,Truck
```

Running the Neural Network

The code listing doesn't include any output messages while it's running, with the exception of the output of the evaluation. I say this because the training data could have 100,000 rows in it, and it's going take a few minutes to run.

Run the class with Run ➪ Run from Eclipse, and it starts to generate the model. After a while, you see the output from the evaluation.

```
Results
======

Correctly Classified Instances        38              100      %
Incorrectly Classified Instances        0                0      %
Kappa statistic                         1
Mean absolute error                     0.0003
Root mean squared error                 0.0004
Relative absolute error                 0.0795 %
Root relative squared error             0.0949 %
Total Number of Instances              38
```

Instances can be easily classified by using the multilayer perceptron `classifyInstance()` method, which takes in a single `Instance` class and outputs a numeric representation of the result. This result corresponds to your output class in the `.arff` training file.

Developing Neural Networks with DeepLearning4J

The Weka framework gives a good working system for creating neural networks. The system that you're using will obviously determine how long the model training will take. From experience I've found that there's a point in the training when Weka starts to struggle. When this happens, I look for the alternatives that I can use. As a Java developer, I use the DeepLearning4J framework; it scales well and also can be used with Spark to let you scale out large datasets across a cluster.

Let's take the vehicle data and use DL4J to create a multilayer perceptron neural network.

Modifying the Data

As Weka has the Arff data file, it knows that the output class is a vehicle type. The training data for DL4J is based on a CSV file, but it requires a numerical output class instead of a text one. So I've changed the output classifications to the following:

VEHICLE CLASS	DL4J NUMERICAL OUTPUT CLASS
Bus	0
Car	1
Truck	2
Bike	3

The data now looks like the following:

```
wheels,chassis,pax,vtype
6,20,39,0
8,23,11,0
5,3,1,1
4,3,4,1
5,3,1,1
4,18,37,0
18,23,2,2
5,4,2,1
1,1,1,3
```

```
18,26,2,2
1,1,1,3
1,1,1,3
1,1,1,3
8,21,28,0
5,4,2,1
```

Viewing Maven Dependencies

In the code repository there is a `pom.xml` file with the required dependencies for DL4J. I'm not using any form of a graphical processor unit (GPU) for the calculations, just the CPU of the machine I'm working on.

```xml
<properties>
    <nd4j.backend>nd4j-native-platform</nd4j.backend>
    <dl4j.version>0.9.1</dl4j.version>
    <nd4j.version>0.9.1</nd4j.version>
</properties>
<dependencies>
    <dependency>
        <groupId>org.nd4j</groupId>
        <artifactId>${nd4j.backend}</artifactId>
        <version>${nd4j.version}</version>
    </dependency>
    <!-- Core DL4J functionality -->
    <dependency>
        <groupId>org.deeplearning4j</groupId>
        <artifactId>deeplearning4j-core</artifactId>
        <version>${dl4j.version}</version>
    </dependency>
    <dependency>
        <groupId>org.deeplearning4j</groupId>
        <artifactId>deeplearning4j-nlp</artifactId>
        <version>${dl4j.version}</version>
    </dependency>
    <dependency>
        <groupId>org.apache.httpcomponents</groupId>
        <artifactId>httpclient</artifactId>
        <version>4.3.5</version>
    </dependency>
    <!-- logging -->
    <dependency>
        <groupId>org.slf4j</groupId>
        <artifactId>slf4j-log4j12</artifactId>
        <version>1.7.13</version>
    </dependency>
    <!-- end logging -->
</dependencies>
```

With that in place, we can now look at the steps to creating the neural network.

Handling the Training Data

As there are various data formats that DL4J can handle, it's important to pick the right one to use for the training data; the CSVRecordReader is used to read in CSV files. The header row will be skipped; this row is text and contains the names of the columns. If this row is read, our model will break when being built.

```
int numLinesToSkip = 1;
String delimiter = ",";
RecordReader recordReader = new CSVRecordReader(numLinesToSkip,
 delimiter);
recordReader.initialize(new FileSplit(
 new File("/path/to/data/ch09/dl4j/")));
```

The next stage is to convert each CSV into an object for DL4J to read. The labeled value we're wanting to predict is the fourth column, the vehicle type, so the labelIndex value is set to 3 (counting from zero). There are four classes in each object, and in this example there are 100,000 rows of data.

I'm going to use 65 percent of the data for training and the remaining 35 percent for evaluating the newly created model. Notice how the dataset is shuffled, so there is some randomness to the training and evaluation data.

```
int labelIndex = 3;
int numClasses = 4;
int batchSize = 100000;
double evalsplit = 0.65;

DataSetIterator iterator = new RecordReaderDataSetIterator(recordReader,
 batchSize,labelIndex,numClasses);

DataSet allData = iterator.next();
allData.shuffle();
SplitTestAndTrain testAndTrain = allData.splitTestAndTrain(evalsplit

DataSet trainingData = testAndTrain.getTrain();
DataSet testData = testAndTrain.getTest();
```

The resulting split leaves us with two new DataSet objects, a collection of training objects, and a collection of evaluation objects. The next step is to normalize the data.

Normalizing Data

In preparing the data for building the neural network, the process of normalization takes places. The aim is to change the values of the vectors to a common scale number but do it without distorting the actual differences of the data being used. The DL4J framework provides a class to do this; in this example, I'm using it to normalize both the training and test vectors.

```
DataNormalization normalizer = new NormalizerStandardize();
normalizer.fit(trainingData);
normalizer.transform(trainingData);
normalizer.transform(testData);
```

Building the Model

Using the `NeuralNetConfiguration.Builder` class, the neural network is constructed. Information about the layers, the hidden nodes, and the output are all created here.

The seed is purely a random number that is used during the model build. We are also required to specify which activation function to use. The Tahn acts very much like a sigmoid function but gives an S curve from -1 to 1, whereas the sigmoid works from 0 to 1.

I'm creating a four-layer network. The input layer has three nodes, the two hidden layers have four nodes for each layer, and the output layer had four nodes, one for each prediction vehicle type. With the `DenseLayer` class, you can see how the `nIn` method handles the input edges and how the `nOut` method sets the output edges for the next connecting layer.

```
final int numInputs = 3;
final int hiddenNodes = 4;
int outputNum = 4;
int iterations = 2000;
long seed = 6;

log.info("Building model....");
MultiLayerConfiguration conf = new NeuralNetConfiguration.Builder()
    .seed(seed)
    .iterations(iterations)
    .activation(Activation.TANH)
    .weightInit(WeightInit.XAVIER)
    .learningRate(0.1)
    .regularization(true).l2(1e-4)
    .list()
    .layer(0, new DenseLayer.Builder().nIn(numInputs).nOut(hiddenNodes)
.build())
    .layer(1, new DenseLayer.Builder().nIn(hiddenNodes).nOut(hiddenNodes)
.build())
    .layer(2, new DenseLayer.Builder().nIn(hiddenNodes).nOut(hiddenNodes)
.build())
    .layer(3, new
```

```
        OutputLayer.Builder(
LossFunctions.LossFunction.NEGATIVELOGLIKELIHOOD)
        .activation(Activation.SOFTMAX)
        .nIn(hiddenNodes).nOut(outputNum).build())
              .backprop(true).pretrain(false)
              .build();
```

With the configuration of the network set, the next stage is to train the model with the configuration and the training dataset that we defined earlier.

```
MultiLayerNetwork model = new MultiLayerNetwork(conf);
model.init();
model.setListeners(new ScoreIterationListener(100));
model.fit(trainingData);
```

Once the model build is complete, it's time to run the evaluation with the other dataset.

Evaluating the Model

The Evaluation class takes an integer with the number of output classes it will evaluate the data against. The model generates an output feature matrix from the test data, and it is evaluated and reported to the console.

```
Evaluation eval = new Evaluation(4);
log.info("Getting evaluation");
INDArray output = model.output(testData.getFeatureMatrix());
log.info("Getting evaluation output");
eval.eval(testData.getLabels(), output);
System.out.println(eval.stats());
```

Saving the Model

The resulting model can be persisted to a file for use again later. To illustrate this, I've added the code to save the model to the filesystem of the local machine.

```
File locationToSave = new File("/path/to/models/basicmlpmodel.zip");
boolean saveUpdater = false;
ModelSerializer.writeModel(model, locationToSave, saveUpdater);
```

The output file type is a zip file. In this example the zip file has two files, a configuration JSON file, which has all the model configuration as created in the code illustrated and a bin file with the generated coefficients.

This model can be loaded and be used to run predictions against new data.

Building and Executing the Program

To enable us to execute the application, I'm going to add the Java execution plugin in to the pom.xml file.

```
<build>
   <plugins>
      <plugin>
         <groupId>org.codehaus.mojo</groupId>
         <artifactId>exec-maven-plugin</artifactId>
         <version>1.2.1</version>
         <executions>
            <execution>
               <goals>
                  <goal>java</goal>
               </goals>
            </execution>
         </executions>
         <configuration>
            <mainClass>mlbook.ch09.ann.dl4j.BasicMLP</mainClass>
         </configuration>
      </plugin>
   </plugins>
</build>
```

The mainClass tag is set to the Java class for our program. To run this, type the following from the command line:

```
$ mvn exec:java -Dexec="mlbook.ch09.ann.dl4j.BasicMLP"
```

You will see the Maven output and, after a few minutes, the results of the evaluation of the neural network model.

```
[INFO] Scanning for projects...
[INFO]
[INFO] ------------------------< mlbook:Chapter9 >--------------------
-------
[INFO] Building Machine Learning:Hands On 2nd Edition -
Chapter 9 - Artificial Neural Networks 1.0-SNAPSHOT
[INFO] ------------------------------[ jar ]-------------------------
-------
[INFO]
[INFO] >>> exec-maven-plugin:1.2.1:java (default-cli) >
validate @ Chapter9 >>>
[INFO]
[INFO] <<< exec-maven-plugin:1.2.1:java (default-cli) < validate @
Chapter9 <<<
[INFO]
[INFO]
[INFO] --- exec-maven-plugin:1.2.1:java (default-cli) @ Chapter9 ---
log4j:WARN No appenders could be found for
```

```
logger (org.nd4j.linalg.factory.Nd4jBackend).
log4j:WARN Please initialize the log4j system properly.
log4j:WARN See http://logging.apache.org/log4j/1.2/faq.html#
noconfig for more info.

Examples labeled as 0 classified by model as 0: 8725 times
Examples labeled as 1 classified by model as 1: 8861 times
Examples labeled as 2 classified by model as 2: 8762 times
Examples labeled as 3 classified by model as 3: 8652 times


=========================Scores=========================================
 # of classes:    4
 Accuracy:        1.0000
 Precision:       1.0000
 Recall:          1.0000
 F1 Score:        1.0000
Precision, recall & F1: macro-averaged (equally weighted avg.
of 4 classes)
========================================================================
[INFO] -----------------------------------------------------------
-------
[INFO] BUILD SUCCESS
[INFO] -----------------------------------------------------------
-------
[INFO] Total time:  03:10 min
[INFO] Finished at: 2019-10-28T14:30:52Z
[INFO] -----------------------------------------------------------
-------
```

Our model took just over three minutes to build, train, and evaluate with 100,000 lines of data. The resulting model was saved to the filesystem, so the model can be reused.

Summary

This is an involved chapter, covering the core concepts of how neural networks actually work. It's worth exploring both the Weka and DeepLearning4J (DL4J) libraries and seeing which one fits the best for your work and projects.

Like I said early in the chapter, it's worth exploring all the other algorithmic options before settling on using a neural network. It's better to have a model that's explainable than not. The black-box nature of these network models makes it incredibly difficult to justify the predictions if anyone questions them.

Care must be taken going forward, especially with live customer data. While it's a computer doing the work and making the predictions, it's still our responsibility to make sure they are fair, are correct, and do not negatively impact another party.

While this chapter has focused on the multilayer perceptron as the neural network of choice, the next two chapters use some of these concepts with convolutional neural networks. With these deep learning algorithms, we can explore large text corpus, images, and video in a machine learning context.

# CHAPTER 10

Machine Learning with Text Documents

The word *document* sounds too formal when you take a moment to consider the amount of text that is stored. That may take the form of a word-processed document, a blog post, an email, a news article, or an academic paper. When you pause to consider the amount of text data held on the Internet and the Web, well, it's a lot, making sense of it is going to take some doing.

Text analysis, and the machine learning from it, is not the easiest thing in the world to do. Documents are messy, there's a fair amount of cleaning to do, and they come in all sorts of different formats, which usually presents challenges too.

For this chapter I will describe various working methods of finding information from text documents but will also cover the steps of getting data ready for analysis. From there I'll show you three methods to learning from your text: TF/IDF, Word2Vec, and using neural networks to generate new text.

As a further study of text analysis, it's worth looking into the more advanced techniques like using Long Short Term Memory (LSTM) for improved results especially in context awareness. Google has designed a neural network architecture called Bidirectional Encoder Representations from Transformers (BERT); a basic overview is available here:

```
https://colab.research.google.com/github/google-research/bert/
blob/master/predicting _ movie _ reviews _ with _ bert _ on _ tf _ hub
.ipynb#scrollTo=hsZvic2YxnTz
```

Preparing Text for Analysis

Let's start at the start; someone, somewhere is going to present you with documents. Previous experience has told me it's going to be in the format you least expect. When we say "text document," some may think of plain text (.txt), while others might think of a Rich Text Format document (.rtf) or even a Microsoft Word document (.doc/.docx). For text analysis we want plain text, so there is usually going to be some data scrubbing to do first.

Apache Tika

If you are totally unsure what kind of document type you are dealing with, then it's worth taking a look at the Apache Tika library to inspect the content metadata.

You can download Tika from http://tika.apache.org/, and it can be used either as a command-line tool or embedded into an application.

Tika isn't just limited to text documents; it can read image, video, sound, email, and other file types. To the see the full list of supported types, please look at the support page on the Apache Tika website.

```
http://tika.apache.org/1.22/formats.html#Full _ list _ of _ Supported _
Formats
```

Downloading Tika

For the command-line examples, I'm going to download the JAR file from the Apache mirror site. To choose your closest mirror, go to the following website and choose from the list:

```
https://www.apache.org/dyn/closer.cgi/tika/tika-app-1.22.jar
```

Once you have saved the JAR file (you may get a security warning about the downloading of a JAR file), you can now reference it from the command line.

A built-in GUI is available to you; from the command line, run the following command.

```
$ java -jar tika-app-1.22.jar
```

Once the GUI (see Figure 10.1) has loaded, find a text file and drag it over the GUI and then drop it there. You'll see the metadata for the file you've dropped.

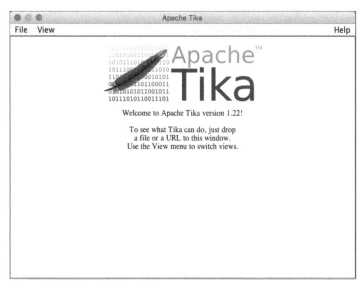

Figure 10.1: Apache Tika GUI

Tika from the Command Line

The same JAR file offers processing from the command line. Using the `--list-parsers` flag, you will see all the supported file types that Tika will read metadata for.

```
$ java -jar tika-app-1.22.jar --list-parsers
Aug 28, 2019 9:27:33 PM
org.apache.tika.config.InitializableProblemHandler
$3 handleInitializableProblem
WARNING: J2KImageReader not loaded. JPEG2000 files will not be
processed.
See https://pdfbox.apache.org/2.0/dependencies.html#jai-image-io
for optional dependencies.
Aug 28, 2019 9:27:33 PM
org.apache.tika.config.InitializableProblemHandler
$3 handleInitializableProblem
WARNING: org.xerial's sqlite-jdbc is not loaded.
Please provide the jar on your classpath to parse sqlite files.
See tika-parsers/pom.xml for the correct version.
    org.apache.tika.parser.AutoDetectParser (Composite Parser):
        org.apache.tika.parser.DefaultParser (Composite Parser):
            org.apache.tika.parser.apple.AppleSingleFileParser
            org.apache.tika.parser.asm.ClassParser
            org.apache.tika.parser.audio.AudioParser
```

```
org.apache.tika.parser.audio.MidiParser
org.apache.tika.parser.chm.ChmParser
org.apache.tika.parser.code.SourceCodeParser
org.apache.tika.parser.crypto.Pkcs7Parser
org.apache.tika.parser.crypto.TSDParser
org.apache.tika.parser.csv.TextAndCSVParser
org.apache.tika.parser.dbf.DBFParser
org.apache.tika.parser.dif.DIFParser
org.apache.tika.parser.dwg.DWGParser
org.apache.tika.parser.epub.EpubParser
org.apache.tika.parser.executable.ExecutableParser
org.apache.tika.parser.feed.FeedParser
org.apache.tika.parser.font.AdobeFontMetricParser
org.apache.tika.parser.font.TrueTypeParser
org.apache.tika.parser.gdal.GDALParser
org.apache.tika.parser.geo.topic.GeoParser
org.apache.tika.parser.geoinfo.GeographicInformationParser
org.apache.tika.parser.grib.GribParser
org.apache.tika.parser.hdf.HDFParser
          .....
org.apache.tika.parser.sas.SAS7BDATParser
org.apache.tika.parser.video.FLVParser
org.apache.tika.parser.wordperfect.QuattroProParser
org.apache.tika.parser.wordperfect.WordPerfectParser
org.apache.tika.parser.xml.DcXMLParser
org.apache.tika.parser.xml.FictionBookParser
org.gagravarr.tika.FlacParser
org.gagravarr.tika.OggParser
org.gagravarr.tika.OpusParser
org.gagravarr.tika.SpeexParser
org.gagravarr.tika.TheoraParser
org.gagravarr.tika.VorbisParser
```

I'm going to use a text file I have on hand; my `nidcclean.txt` file contains all the conference talk descriptions from a local developer conference.

First, I want some metadata on the file.

```
$ java -jar tika-app-1.22.jar -m ~/nidcclean.txt
Aug 28, 2019 9:36:23 PM org.apache.tika.config.Initializable
ProblemHandler$3 handleInitializableProblem
WARNING: J2KImageReader not loaded. JPEG2000 files will not be
processed.
See https://pdfbox.apache.org/2.0/dependencies.html#jai-image-io
for optional dependencies.
Aug 28, 2019 9:36:23 PM org.apache.tika.config.Initializable
ProblemHandler$3 handleInitializableProblem
WARNING: org.xerial's sqlite-jdbc is not loaded.
Please provide the jar on your classpath to parse sqlite files.
See tika-parsers/pom.xml for the correct version.
Content-Encoding: UTF-8
```

```
Content-Length: 38222
Content-Type: text/plain; charset=UTF-8
X-Parsed-By: org.apache.tika.parser.DefaultParser
X-Parsed-By: org.apache.tika.parser.csv.TextAndCSVParser
resourceName: nidcclean.txt
```

The -m option flag will output the meta data. Supposing I want to dump out of the file to XML instead, that can be done from the command line too.

```
$ java -jar tika-app-1.22.jar -x ~/nidcclean.txt
Aug 28, 2019 9:44:33 PM org.apache.tika.config.Initializable
ProblemHandler$3 handleInitializableProblem
WARNING: J2KImageReader not loaded. JPEG2000 files will not be
processed.
See https://pdfbox.apache.org/2.0/dependencies.html#jai-image-io
for optional dependencies.

Aug 28, 2019 9:44:33 PM org.apache.tika.config.Initializable
ProblemHandler$3 handleInitializableProblem
WARNING: org.xerial's sqlite-jdbc is not loaded.
Please provide the jar on your classpath to parse sqlite files.
See tika-parsers/pom.xml for the correct version.
<?xml version="1.0" encoding="UTF-8"?>
<html xmlns="http://www.w3.org/1999/xhtml">
<head>
<meta name="X-Parsed-By"
content="org.apache.tika.parser.DefaultParser"/>
<meta name="X-Parsed-By" content="org.apache.tika.parser.csv.
TextAndCSVParser"/>
<meta name="Content-Encoding" content="UTF-8"/>
<meta name="resourceName" content="nidcclean.txt"/>
<meta name="Content-Length" content="38222"/>
<meta name="Content-Type" content="text/plain; charset=UTF-8"/>
<title/>
</head>
<body><p>visualising biological information is challenging at the best
of times at axial3d we accelerate that understanding by providing
machine learning ml backed annotations
....
 aimed at beginners mostly because i am one tootestcontainers is an
open source library that allows you to containerise your external
resource dependencies like databases web browsers or anything that can
run in a docker container!by making use of testcontainers we can to
develop and run our tests easier in a more productionlike environment
with only docker as a prerequisite</p>
</body></html>
```

With Tika you can safely do text extraction. Let's look at a résumé for an example (also joyously called a *curriculum vitae* if you're in the United Kingdom). My file is in PDF format, but I want to quickly extract the text.

```
$ java -jar tika-app-1.22.jar -t ~/Documents/JasonBellCV2018.pdf
Aug 28, 2019 9:48:39 PM    org.apache.tika.config.Initializable
ProblemHandler$3 handleInitializableProblem
WARNING: J2KImageReader not loaded. JPEG2000 files will not be
processed.
See https://pdfbox.apache.org/2.0/dependencies.html#jai-image-io
for optional dependencies.
Aug 28, 2019 9:48:39 PM org.apache.tika.config.Initializable
ProblemHandler$3 handleInitializableProblem
WARNING: org.xerial's sqlite-jdbc is not loaded.
Please provide the jar on your classpath to parse sqlite files.
See tika-parsers/pom.xml for the correct version.

Profile
Highly proficient machine learning and data engineer with experience
in building and
maintaining high volume data pipelines, realtime stream
processing systems and
machine learning solutions for a variety of customers.
Experienced in data cleaning
and preparation for use within data intensive systems.

Comfortable in both the development process and the customer facing/
communication process and is involved in the software industry
as a respected voice in
the data community and is asked to speak at various events
on Artificial Intelligence,
Machine Learning and anything to do with data.
```

The -t option will take the file content and output the text. You can direct that to a file if you want.

Tika Within an Application

In the code repository there is a code folder for this chapter. The Apache Tika libraries are available in two versions depending on the use you need. The core libraries contain everything for extracting metadata; if you want to extract text and conversions, you will need the parsers release.

```
package mlbook.ch10.tika;
import org.apache.tika.exception.TikaException;
import org.apache.tika.metadata.Metadata;
import org.apache.tika.parser.AutoDetectParser;
import org.apache.tika.parser.pdf.PDFParser;
import org.apache.tika.sax.BodyContentHandler;
import org.xml.sax.SAXException;
import java.io.BufferedInputStream;
import java.io.FileInputStream;
import java.io.IOException;
import java.io.InputStream;
```

```
public class TextExtraction {
    public String toPlainText(String filename) {
        BodyContentHandler handler = new BodyContentHandler();
        AutoDetectParser parser = new AutoDetectParser();
        Metadata metadata = new Metadata();
        String output = "";
        try {
            InputStream stream = new BufferedInputStream(
new FileInputStream(filename));
            parser.parse(stream, handler, metadata);
            output =  handler.toString();
        } catch (IOException e) {
            e.printStackTrace();
        } catch (TikaException e) {
            e.printStackTrace();
        } catch (SAXException e) {
            e.printStackTrace();
        }
        return output;
    }

    public static void main(String[] args) {
        TextExtraction t = new TextExtraction();
        String output = t.toPlainText("/path/to/your/file.pdf");
        System.out.println(output);
    }
}
```

Cleaning the Text Data

When presented with text data, you will usually need to do some form of cleaning. What needs cleaning is down to the specification of the project, but you'll come across a few commonalities.

1. Extract text from document (if not a plain text document).
2. Convert all words to lowercase.
3. Remove any punctuation.
4. Remove stopwords.

Convert Words to Lowercase

Lowercase and uppercase characters will be treated as separate words in any analysis, so it's important to convert the entire text to one or the other. Consider the following sentence:

```
You are worthy, you are family. Focus on what you have to be
grateful for.
```

The words *You* and *you* would be treated as two separate words by any algorithm, and this would influence any weights or scorings.

In Java the .toLowerCase method will transform the string to lowercase.

```
public static String convertToLowerCase(String in) {
    return in.toLowerCase();
}
```

Using Clojure the lowercase function in the clojure.string library will do the same.

```
user=> (clojure.string/lower-case "You are worthy, you are family.
 Focus on what you have to be grateful for.")
"you are worthy, you are family. focus on what you have to be
 grateful for."
```

Remove Punctuation

In the same way that words with uppercase and lowercase instances will be recognized twice, the same applies to words with punctuation attached.

Let's get back to the example text in its current state.

```
"you are worthy, you are family. focus on what you have to be
 grateful for."
```

The words *worthy*, *family*, and *for* have punctuation in the way of commas and full stops. If the punctuation is not removed, then these also become classed as separate words in any analysis. If *family* and *family.* were in the same paragraph, then there would be two instances recorded.

Using the regular expressions package in Java gives us a usable solution. The Pattern class takes the actual regular expression we want to use (\w is a word class that is any word containing ASCII letters, numbers, or an underscore [_] character). The Matcher class then gives the results of the regular expression applied to the input string. I'm using a StringBuilder to create an output string that can be used.

```
public String removePunctuation(String in) {
    String patternString = "[\\w]+";

    Pattern pattern = Pattern.compile(patternString);
    Matcher matcher = pattern.matcher(convertToLowerCase(in));
    StringBuilder sb = new StringBuilder();

    while(matcher.find()) {
        sb.append(matcher.group() + " ");
    }
    return sb.toString().trim();
}
```

Clojure uses the same Java function, but it's wrapped in a handy sequence so you can iterate through the results.

```
user=> (re-seq #"[\w]+" "you are worthy, you are family. focus on
what you have to be grateful for.")
("you" "are" "worthy" "you" "are" "family" "focus" "on" "what"
"you" "have" "to" "be" "grateful" "for")
```

Stopwords

It's worth pausing for a moment to consider stopwords. In most cases, there are common words that you want to remove so as not to get in the way of analysis. For a list of common stopwords, this list is a good starting point.

```
$ cat stopwords.txt
a about above after again against all am an and any are as at be
because been before being below between both but by can did do does
doing don down during each few for from further had has have having he
her here hers herself him himself his how i if in into is it its itself
just me more most my myself no nor not now of off on once only or other
our ours ourselves out over own s same she should so some such t than
that the their theirs them themselves then there these they this those
through to too under until up very was we were what when where which
while who whom why will with you your yours yourself yourselves
```

You may find that a list of common stopwords is not enough. The domain that you work in may have common words that, while not commonly used words generally, are distorting the results of your analysis. At this point, you have a decision to make: either append your domain-level words to the same stopword file or have a separate file of domain-specific words.

With the use of the Java Collections API, there is the option to use the Stream API to convert a string to an `ArrayList`. Using the `removeAll` method, the stopwords can be passed in. The resulting string is the content with the stop words removed.

First load in your text file and then convert it to lowercase.

```
    rawtext = new String(Files.readAllBytes(
Paths.get("yourdatafile.txt")));
    rawtext = rawtext.toLowerCase();
```

Next, load in the stopwords.

```
    stopwords = Files.readAllLines(Paths.get("stopwords.txt"));
```

With the raw text loaded and converted to lowercase, the stopwords are also loaded. The next step is to convert the raw text string to an array and remove all the occurrences that appear in the stopwords. Lastly, the outgoing string is joined by a space, so you are left with one cleaned string.

```
public String removeAll() {
    ArrayList<String> importtext =
      Stream.of(rawtext.split(" "))
            .collect(Collectors.toCollection(ArrayList<String>::new));
    importtext.removeAll(stopwords);
    return importtext.stream().collect(Collectors.joining(" "));
}
```

Stemming

Though not essential, it can be useful to stem phrases down to their root form. Word derivations are common, and for some analysis converting all those different forms to a root word can be useful.

If we look at the word *like*, for example, you may come across instances in your corpus of *likes*, *likely*, *liked*, and *liking*. When a stemming function is applied, then you'd return with the root of the word, *like*.

Care must be given with stemming text as there is a risk of over stemming, where the text is cut back to a root that actually may have two different meanings and your analysis will then miss out on the context.

The Apache OpenNLP project (`https://opennlp.apache.org`) provides a number of stemming applications for your text. It also offers language detection, tagging, and other tools.

N-grams

N-grams are sequences of words and are often used in natural language processing.

```
you are worthy you are family focus on what you have to be grateful for
```

The previous line has 15 words, so it's a 15-gram. If this is split into more sensible two- or three-word n-grams, it may be possible to predict the next word groupings based on the n-gram sequences.

A two-gram sequence of the sentence would look something like this:

```
(you are), (are worthy), (worthy you), (you are), (are family)....
```

And so on. Interestingly, there are two 2-gram sequences of *(you are)* with two following word patterns, the *worthy* and the word *family*.

The three-word n-gram sequence would look like this:

```
(you are worthy), (are worthy you), (worthy you are), (you are
family)....
```

Having n-gram sequences of words can be used against algorithms such as Term Frequency/Inverse Document Frequency, which will be covered later in this chapter. Each sequence in the n-gram can then be used as a term. Sometimes this will give more meaningful scorings than single words; it also means that we are scoring within the context of the corpus text.

TF/IDF

One useful technique is to find out how important a word or phrase is within a corpus of text or a collection of documents. Term Frequency/Inverse Document Frequency (TF/IDF) is a method of giving a numerical value to the importance of a word. TF/IDF is used widely within recommendation systems, and it's quite easy to implement.

To give you an idea of how it works, let's work through some sample code to build a TF/IDF algorithm in Java.

Loading the Documents

Before any calculations can be done, we need to load the documents into the application. The documents are loaded and split on the space character; each word is then added to a List collection and returned. The reason for using a List of words is simple; it will be easier to iterate and count the word frequencies.

I'm assuming that the document is clean, as in it has been converted to lowercase and the punctation has already been removed.

For every document that you want included in your document set, you would execute this step and load the document.

```java
public List<String> loadDocToStrings(String filepath) {
    List<String> words = new ArrayList<String>();
    try {
        File file = new File(filepath);
        BufferedReader br = new BufferedReader(new FileReader(file));
        String s;
        while ((s = br.readLine()) != null) {
            String[] ws = s.split(" ");
            for (int i = 0; i < ws.length; i++) {
                words.add(ws[i]);
            }
        }
    } catch(IOException e) {
        e.printStackTrace();
    }
    return words;
}
```

Finally, with all the documents loaded into separate `List` objects, a final `List` of documents is then created.

```
List<List<String>> allDocuments = Arrays.asList(wordDoc1, wordDoc2,
wordDoc3, wordDoc4, wordDoc5);
```

Calculating the Term Frequency

The *term frequency* is the number of times a phrase is in the document; in this example we're using an iterator over the list of words and seeing whether the term matches.

If there is a match, then the count is increased by one.

```
public double getTermFrequency(List<String> doc, String term) {
    double result = 0;
    for (String word : doc) {
        if (term.equalsIgnoreCase(word))
            result++;
    }
    return result / doc.size();
}
```

The final step of calculating the term frequency is to divide the number of occurrences against the size of the document.

Calculating the Inverse Document Frequency

The inverse document frequency is calculated against all the documents in the corpus, this is why the collection of word lists was created when the files were loaded. This measure provides us with an indication of how common, or rare, the term is against the complete corpus of documents.

Similarly, the term frequency calculation is counted against all the documents, iterating through each word.

```
public double getInverseDocumentFrequency(List<List<String>>
allDocuments, String term) {
    double wordOccurances = 0;
    for (List<String> document : allDocuments) {
        for (String word : document) {
            if (term.equalsIgnoreCase(word)) {
                wordOccurances++;
                break;
            }
        }
    }
    return Math.log(allDocuments.size() / wordOccurances);
}
```

The result is the number of documents divided by the number of times the term was found in those documents; the logarithm of the quotient is the value passed back.

Computing the TF/IDF Score

The final step is to compute the TF/IDF score. This is done by simply multiplying the result of the term frequency with the score of the inverse document frequency.

```
    public double computeTfIdf(List<String> doc, List<List<String>>
docs, String term) {
        return getTermFrequency(doc, term) *
getInverseDocumentFrequency(docs, term);
    }
```

Assuming we are looking for the score for the term *dapibus*, the document with the term frequency we want to calculate, along with the entire corpus and the term, is passed into the computeTfIdf method.

```
double score = tfidf.computeTfIdf(wordDoc4, allDocuments, "dapibus");
```

If the term were to appear in more documents, then the score would begin to reach the value of 1. In this instance, *dapibus* does not appear in the corpus often and has little weight in the scoring.

```
Term Frequency for dapibus in wordDoc4 = 0.009009009009009009
Inverse Doc Frequency for dapibus = 1.6094379124341003
TF-IDF score for the word: dapibus = 0.014499440652559462
```

Reviewing the Final Code Listing

Listing 10.1 is the full code for the basic TF/IDF algorithm that has been explained. Other implementations exist in both Spark and DeepLearning4J, which will give you better control and handling of larger corpus datasets.

Listing 10.1: Basic TF/IDF Algorithm

```
package mlbook.ch10.tfidf;

import java.io.BufferedReader;
import java.io.File;
import java.io.FileReader;
import java.io.IOException;
import java.util.ArrayList;
import java.util.Arrays;
import java.util.List;
```

```java
public class TFIDFExample {
    public double getTermFrequency(List<String> doc, String term) {
        double result = 0;
        for (String word : doc) {
            if (term.equalsIgnoreCase(word))
                result++;
        }
        return result / doc.size();
    }

    public double getInverseDocumentFrequency(List<List<String>>
allDocuments, String term) {
        double wordOccurances = 0;
        for (List<String> document : allDocuments) {
            for (String word : document) {
                if (term.equalsIgnoreCase(word)) {
                    wordOccurances++;
                    break;
                }
            }
        }
        return Math.log(allDocuments.size() / wordOccurances);
    }

    public double computeTfIdf(List<String> doc, List<List<String>>
docs, String term) {
        return getTermFrequency(doc, term) *
getInverseDocumentFrequency(docs, term);
    }

    public List<String> loadDocToStrings(String filepath) {
        List<String> words = new ArrayList<String>();
        try {
            File file = new File(filepath);
            BufferedReader br = new BufferedReader(new FileReader(file));
            String s;
            while ((s = br.readLine()) != null) {
                String[] ws = s.split(" ");
                for (int i = 0; i < ws.length; i++) {
                    words.add(ws[i]);
                }
            }
        } catch(IOException e) {
            e.printStackTrace();
        }
        return words;
    }
```

```
public static void main(String[] args) {
    String docspath = "/path/to/data/ch10";
    TFIDFExample tfidf = new TFIDFExample();

    List<String> wordDoc1 = tfidf.loadDocToStrings(docspath +
"/doc1.txt");
    List<String> wordDoc2 = tfidf.loadDocToStrings(docspath +
"/doc2.txt");
    List<String> wordDoc3 = tfidf.loadDocToStrings(docspath +
"/doc3.txt");
    List<String> wordDoc4 = tfidf.loadDocToStrings(docspath +
"/doc4.txt");
    List<String> wordDoc5 = tfidf.loadDocToStrings(docspath +
"/doc5.txt");

    List<List<String>> allDocuments = Arrays.asList(wordDoc1,
wordDoc2, wordDoc3, wordDoc4, wordDoc5);

    double score = tfidf.computeTfIdf(wordDoc4, allDocuments,
"dapibus");
    System.out.println("TF-IDF score for the word: dapibus = " +
score);
    }
}
```

Word2Vec

The Word2Vec algorithm was developed by Google. It comprises a neural network of two layers. With a large corpus of text you can achieve some very accurate vector results. Groups of words will appear closer within the vectors. It's not just limited to text; you can use this method on pretty much anything where patterns of associations would occur; this might be personality scorings in a social network or what kind of music you are into.

The Word2Vec algorithm is based on vectors, called *neural word embeddings*, representing a word with numbers. Word2Vec trains words against other words in the input text. This is done in one of two ways, either using a continuous bag of words (CBOW), which is a context of words to predict a target word, or using skip grams, which takes a word and predicts a context of words.

Words are read in the vector one at a time and then scanned within a certain range of words; these skip-grams are an n-gram with items dropped. During the training, the vector contains the context of each word and the similarity against other words in the vector space.

In this section, I will outline how to construct a Word2Vec implementation using DeepLearning4J. First I'll explain what's going on in the code, and then you'll be able to see the full code listing at the end.

Loading the Raw Text Data

The first job is to load in the raw text. The `LineSentenceIterator` will give an iterator and preprocess the file with the `preProcess` method within the inner class. Here I'm going to convert the string to lowercase.

```
public SentenceIterator createSentenceIterator(String filepath) {
        SentenceIterator iter = new LineSentenceIterator(new
File(filepath));
        iter.setPreProcessor(new SentencePreProcessor() {
            public String preProcess(String sentence) {
                return sentence.toLowerCase();
            }
        });
        return iter;
    }
```

Tokenizing the Strings

The next job is to tokenize the strings. For the basic one-word tokenizer splitting on a whitespace, the `CommonPreprocessor` will work fine for us.

```
public TokenizerFactory createTokenizer() {

        TokenizerFactory t = new DefaultTokenizerFactory();
        t.setTokenPreProcessor(new CommonPreprocessor());
        return t;
    }
```

Creating the Model

With our sentence iterator and tokenizer created, we can now build the model. DeepLearning4J provides a convenient Word2Vec model that we can implement.

```
public Word2Vec createWord2VecModel(SentenceIterator iter,
TokenizerFactory t) {
        Word2Vec vec = new Word2Vec.Builder()
                .minWordFrequency(5)
                .layerSize(100)
                .seed(42)
                .windowSize(5)
                .iterate(iter)
                .tokenizerFactory(t)
                .build();
        vec.fit();
        return vec;
    }
```

There are some parameters that are set. The `minimumWordFrequency` value is the number of times the word must appear in the corpus. The number of features in a vector is set with the `layerSize` method; in our example, there are 100 features in this vector space.

The final step in the model is `vec.fit()` where the training begins. When finished, it returns the model.

Evaluating the Model

The feature vector values for the model are written to disk. It is possible to load and update the model when new data is added.

```
public void evaluateModel(Word2Vec vec) {
    try {
        System.out.println("Serializing the model to disk.");
        WordVectorSerializer.writeWordVectors(vec,
"word2vecoutput.txt");
    } catch(IOException e) {
        e.printStackTrace();
    }

}
```

So, how does our new model look? Let's run some basic tests and see what the output looks like. First let's look at the word associations; I want to know the words that are nearest to the word *data*.

```
Collection<String> lst = vec.wordsNearest("data", 10);
```

The `wordsNearest` method takes the word we want the associations for and how many words to return. That will return a collection of strings that I can iterate over and process.

```
[in, machine, are, learning, over, a, can, out, it, good]
```

Note there are a few common words; it's a good idea to strip these words out prior to training.

Now I'd like to see the closeness between the word *machine* and the words *data*, *retail*, and *games*. For this I need to use the `similarity()` method. It takes two strings that are words from the corpus, and it outputs a number indicating the cosine similarity.

```
System.out.println("Similarity score for data:machine - " +
vec.similarity("data", "machine"));
System.out.println("Similarity score for retail:machine - " +
vec.similarity("retail", "machine"));
```

```
System.out.println("Similarity score for games:machine - " +
vec.similarity("games", "machine"));

Similarity score for data:machine - 0.9937633872032166
Similarity score for retail:machine - 0.6593816876411438
Similarity score for games:machine - 0.9365512132644653
```

The higher the number value, the "closer" in similarity that word is to the target word within the corpus. In the previous example, we see that *data* is very close to the word *machine* as is *games*.

Reviewing the Final Code

Listing 10.2 is the final, complete code for our Word2Vec model. Word2Vec works best with large datasets; the smaller the corpus of text, the less quality you'll see in your associated word collections.

Listing 10.2: Word2Vecmodel

```
package mlbook.ch10.word2vec;
import org.deeplearning4j.models.embeddings.loader.WordVectorSerializer;
import org.deeplearning4j.models.word2vec.Word2Vec;
import org.deeplearning4j.text.sentenceiterator.LineSentenceIterator;
import org.deeplearning4j.text.sentenceiterator.SentenceIterator;
import org.deeplearning4j.text.sentenceiterator.SentencePreProcessor;
import org.deeplearning4j.text.tokenization.tokenizer.preprocessor.
CommonPreprocessor;
import org.deeplearning4j.text.tokenization.tokenizerfactory.
DefaultTokenizerFactory;
import org.deeplearning4j.text.tokenization.tokenizerfactory.
TokenizerFactory;
import java.io.File;
import java.io.IOException;
import java.util.Collection;

public class Word2VecExample {

    public Word2VecExample() {
        System.out.println("Creating sentence iterator");
        SentenceIterator iter = createSentenceIterator("/path/to/data/
ch10/ word2vec_test.txt");

        System.out.println("Creating tokenizer.");
        TokenizerFactory t = createTokenizer();
        System.out.println("Creating word2vec model.");
        Word2Vec vec = createWord2VecModel(iter, t);
        System.out.println("Evaluating the model.");
        evaluateModel(vec);
    }
```

```java
    public Word2Vec createWord2VecModel(SentenceIterator iter,
TokenizerFactory t) {
        Word2Vec vec = new Word2Vec.Builder()
                .minWordFrequency(5)
                .layerSize(100)
                .seed(42)
                .windowSize(5)
                .iterate(iter)
                .tokenizerFactory(t)
                .build();
        vec.fit();
        return vec;
    }


    public SentenceIterator createSentenceIterator(String filepath) {
        SentenceIterator iter = new LineSentenceIterator(new
File(filepath));
        iter.setPreProcessor(new SentencePreProcessor() {
            public String preProcess(String sentence) {
                return sentence.toLowerCase();
            }
        });
        return iter;
    }


    public TokenizerFactory createTokenizer() {
        // Split on white spaces in the line to get words
        TokenizerFactory t = new DefaultTokenizerFactory();
        t.setTokenPreProcessor(new CommonPreprocessor());
        return t;
    }


    public void evaluateModel(Word2Vec vec) {
        try {
            System.out.println("Serializing the model to disk.");
            WordVectorSerializer.writeWordVectors(vec,
"word2vecoutput.txt");
        } catch(IOException e) {
            e.printStackTrace();
        }

        System.out.println("Finding words nearest the word 'machine'.");
        Collection<String> lst = vec.wordsNearest("retail", 10);
        System.out.println(lst);
System.out.println("Similarity score for data:machine - " +
vec.similarity("data", "machine"));
System.out.println("Similarity score for retail:machine - " +
vec.similarity("retail", "machine"));
System.out.println("Similarity score for games:machine - " +
vec.similarity("games", "machine"));
    }
```

```
    public static void main(String[] args) {
        Word2VecExample w2ve = new Word2VecExample();
    }
}
```

Basic Sentiment Analysis

There is always a lot of interest around sentiment analysis, especially with the amount of data generated by social media. My first investigations into Big Data were around large volumes of Twitter data from things like the MTV Music Awards. Some of the techniques I used then I still use now, because they are simple and work nicely. It also means they are easy for anyone else to pick up.

The basic process works like this:

- Load in a set of positive words.
- Load in a set of negative words.
- Load in a set of sentences to measure the sentiment of.
- For each sentence, split on the space character so there is a collection of words.
- Set the score variable to zero.
- Iterate the collection, and for each positive word found, add one to the score; if a negative word is found, subtract one from the score.

While not overly exciting as machine learning or processing goes, it works and can also be implemented in most languages easily. Let's take a look at a basic Java implementation.

Loading Positive and Negative Words

The loadWords method loads in a text file with either positive or negative words. As the file has comments that start with a semicolon character (;), we need to ignore these lines and just add the words to the Set.

```
    public Set<String> loadWords(String filepath) {
        Set<String> words = new HashSet<String>();
        try {
            File file = new File(filepath);
            BufferedReader br = new BufferedReader(new FileReader(file));
            String s;
            while ((s = br.readLine()) != null) {
                if(!s.startsWith(";")) {
```

```
                    words.add(s);
                }
            }
        } catch(IOException e) {
            e.printStackTrace();
        }
        return words;
    }
```

This is done for both positive and negative word sets.

Loading Sentences

The sentences that we want to measure sentiment against are loaded in and added to a `List` of `Strings`. In this example, the assumption is that the data is cleaned, converted to lowercase, and has the punctuation removed.

```
public List<String> loadSentences(String filepath) {
    List<String> sentences = new ArrayList<String>();
    try {
        File file = new File(filepath);
        BufferedReader br = new BufferedReader(new FileReader(file));
        String s;
        while ((s = br.readLine()) != null) {
            sentences.add(s);
        }
    } catch(IOException e) {
        e.printStackTrace();
    }
    return sentences;
}
```

Calculating the Sentiment Score

Now let's talk about the main part of the program, the sentiment score itself. The `calculateSentimentScore` takes three parameters: the sentence to be scored, the positive word set, and the negative word set. The sentence is split by the space character, which gives us a primitive `String array` (`String[]`).

```
public int calculateSentimentScore(String sentence, Set<String>
pwords, Set<String> nwords) {
    int score = 0;
    String[] words = sentence.split(" ");
    for (int i = 0; i < words.length; i++) {
        if(pwords.contains(words[i])) {
            System.out.println("Contains the positive word: " +
```

```
words[i]);
                    score = score + 1;
            } else if (nwords.contains(words[i])) {
                System.out.println("Contains the negative word: " +
words[i]);
                    score = score - 1;
            }
        }
        return score;
    }
```

The score is calculated by iterating the sentence string array. If the word in the loop is within the positive word set, we add one to the count. If it appears in the negative set, then we subtract one from the score. Once all the words have been iterated, the score is returned.

Reviewing the Final Code

Listing 10.3 is the final code for the sentiment analysis program. The sentences and positive and negative word sets are loaded and then processed.

Listing 10.3: Sentiment Analysis Program

```
package mlbook.ch10.sentiment;

import java.io.BufferedReader;
import java.io.File;
import java.io.FileReader;
import java.io.IOException;
import java.util.*;

public class BasicSentimentAnalysis {
    public BasicSentimentAnalysis() {}

    public void runSentimentAnalysis(List<String> sentences) {
        Set<String> pwords = loadWords("/path/to/data/ch10/
sentiment/positive-words.txt");
        Set<String> nwords = loadWords("/path/to/data/ch10/
sentiment/negative-words.txt");
        for(String s : sentences) {
            System.out.println("Sentence: " + s);
            System.out.println("Score: " + calculateSentimentScore(s,
pwords, nwords));
            System.out.println("*******");
        }
    }

    public int calculateSentimentScore(String sentence, Set<String>
pwords,     Set<String> nwords) {
```

```java
        int score = 0;
        String[] words = sentence.split(" ");
        for (int i = 0; i < words.length; i++) {
            if(pwords.contains(words[i])) {
                System.out.println("Contains the positive word: " +
words[i]);
                score = score + 1;
            } else if (nwords.contains(words[i])) {
                System.out.println("Contains the negative word: " +
words[i]);
                score = score - 1;
            }
        }
        return score;
    }

    public List<String> loadSentences(String filepath) {
        List<String> sentences = new ArrayList<String>();
        try {
            File file = new File(filepath);
            BufferedReader br = new BufferedReader(new FileReader(file));
            String s;
            while ((s = br.readLine()) != null) {
                sentences.add(s);
            }
        } catch(IOException e) {
            e.printStackTrace();
        }
        return sentences;
    }

    public Set<String> loadWords(String filepath) {
        Set<String> words = new HashSet<String>();
        try {
            File file = new File(filepath);
            BufferedReader br = new BufferedReader(new FileReader(file));
            String s;
            while ((s = br.readLine()) != null) {
                if(!s.startsWith(";")) {
                    words.add(s);
                }
            }
        } catch(IOException e) {
            e.printStackTrace();
        }
        return words;
    }

    public static void main(String[] args) {
        BasicSentimentAnalysis bsa = new BasicSentimentAnalysis();
```

```
    List<String> sentences = bsa.loadSentences("/path/to/data/
ch10/sentiment/sentences.txt");
    bsa.runSentimentAnalysis(sentences);
  }
}
```

Performing a Test Run

Let's give the sentiment analysis a run and see how it's working. In the data directory there are some sample sentences.

```
i loved receiving the gifts from you it was like it was my birthday
i hated that movie
this is the best meal i've ever had
this is the worst meal i've ever had
```

Now let's run those sentences through the program and see how the output looks.

```
Sentence: i loved receiving the gifts from you it was like it was my
birthday
Contains the positive word: loved
Contains the positive word: like
Score: 2
*******
Sentence: i hated that movie
Contains the negative word: hated
Score: -1
*******
Sentence: this is the best meal i've ever had
Contains the positive word: best
Score: 1
*******
Sentence: this is the worst meal i've ever had
Contains the negative word: worst
Score: -1
*******
```

I've added some verbose statements so you can see where the scoring is happening. In most cases, you wouldn't be overly interested in knowing which words were triggering the scores but just the final sentiment score.

Further Development

There is plenty of scope to improve on the basic sentiment analysis code, especially where the data is coming from. For me, the next obvious point of call would be the source data. In Appendix B, there are instructions on how to set up a Twitter app through the developer account. With that in place, you can start to pull public tweets and apply sentiment analysis.

The same functions could also be used with Kafka and Spark to allow sentiment scoring at volume and velocity.

Summary

This chapter dealt with various considerations when working with text data. It covered the acquisition, conversion, and cleanup data as well as the analysis of it. It also covered how to find the importance of words with Term Frequency/Inverse Document Frequency, word groupings with Word2Vec, and sentiment scoring.

With the text dealt with, the next logical step is to look at how to process images.

Machine Learning with Images

So far in this book the training and classification of information has been based around either datasets of numbers or, as in Chapter 10, text.

In this chapter, we'll take a brief look at image processing and classification, starting with using a basic neural network and then extending that knowledge to use convolutional neural networks for image classification.

Over the last few years there have been huge leaps forward in image processing with machine learning. The addition of graphic processing units (GPUs) will speed up the training of models. To get an idea of how good things have gotten, take a look at the website This Person Does Not Exist (`https://thispersondoes-notexist.com`). Using the StyleGAN model developed by Nvidia, each of the images is generated and is not a real person, but they look alarmingly realistic!

What Is an Image?

In its basic form, a computer-based image is a grid of numbers. Each "square" is called a *pixel*. Figure 11.1 is an example of an 8 pixel by 8–pixel image.

Not overly artistic I agree, but it's a starting point. Let's assume this is an image of two colors: black and white. When there is a pixel colored black, then it's given the value of one, and all the others are zero. From a numeric point of view, our image looks like Figure 11.2.

What we have is a 1 bitmap image representation. Each bit represents the color, black or white.

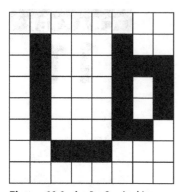

Figure 11.1: An 8 x 8–pixel image

0	0	0	0	0	0	0	0
0	1	0	0	0	1	0	0
0	1	0	0	0	1	1	1
0	1	0	0	0	1	0	1
0	1	0	0	0	1	0	1
0	1	0	0	0	1	1	0
0	0	1	1	1	0	0	0
0	0	0	0	0	0	0	0

Figure 11.2: Numeric image of an 8 x 8–pixel

Introducing Color Depth

The more bits available, the more information you can store in the image. Table 11.1 shows the image information that can be handled depending on the image depth; the larger the depth, the more colors that can be introduced.

Table 11.1: Image Color Depth

COLOR DEPTH IN BITS	NUMBER OF COLORS	EXAMPLE BINARY
1	2	0,1
2	4	00,01,10,11
3	8	000,001,011, etc.
4	14	0000,0001,0011, etc.
8	256	01001001, 11100011, etc.
24	16,777,216	010110101010011011111101

Even at 24 bits, an image, such as Figure 11.3, is just a collection of numbers in grid form; there's just more information going on. The image illustrated in Figure 11.1 is 8 × 8 pixels and has only two colors (black or white, zero or one). On the other hand, the sunflower was a 24-bit color image and has 15,360,000 bits of information. In the context of machine learning, it may be prudent to reduce the color depth to speed up training; reducing the image size will help too.

Figure 11.3: 24-bit image

Images in Machine Learning

As you are aware from reading this book and working through the examples, most of what we are doing with machine learning is feeding information, usually in number form, and finding patterns that can then be defined into models.

So, if we can convert image data into a grid of numbers, what we are left with is a matrix grid of numbers that a machine learning algorithm can train against.

You will find that most of the machine learning examples use a fairly small grid of numbers; images that are 16 × 16 and 28 × 28 pixels are used a lot. The reason for this is processing time; if an image is too large, then it will take a long time, depending on machine performance, to convert and process the image. When you are dealing with tens of thousands of images, these processes can take hours.

In most cases, it is prudent to be prepared and size the images small enough for processing. It's also important to ensure that the image set is using the same height and width in pixels.

With all this in mind, let's take a look at processing images with a basic neural network.

Basic Classification with Neural Networks

We've already covered how neural networks function in Chapter 9; the same multilayer perceptron can be applied to images. The work is in converting the image information into numerical form.

The DeepLearning4J framework provides several file input classes to use, so it's possible to read in a directory of images and convert them to be ready for training.

If you've spent any time looking in books or across the Internet for machine learning examples when it comes to images, then you may have seen the Modified National Institute of Standards and Technology database (MNIST database). It's a large database of handwritten digits, and it's widely used for training image processing systems. The original database has 60,000 images for training and 10,000 for testing and evaluation.

To help even further, they are sized as 28 × 28 pixels and are black and white using a grayscale palette.

For the following example using a multilayer neural network, I'll use this dataset. You don't have to download the actual images yourself; within the DeepLearning4J libraries are helper functions to do that for you.

Basic Settings

The image size, we've established, is 28 × 28 pixels. There are 10 output classes 0–9. Training will happen in batches of 64 images. The rate is the learning rate of the multilayer perceptron.

```
int imageHeight = 28;
int imageWidth = 28;
int outputClasses = 10;
int batchSize = 64;
int randomSeed = 123;
int epochs = 15;
double rate = 0.0015;
```

Loading the MNIST Images

We need to define two datasets, one for the training data and another for the test data to evaluate the model. The DeepLearning4J framework provides some helper classes to load in the MNIST image data set; this means we're not wasting time having to reinvent the wheel.

`MnistDataSetIterator` takes the batch size (64) and a Boolean value to indicate whether we are dealing with a training or a test data set. The random seed value is used so we get some shuffling going on in the dataset and not the same files back each time.

```
DataSetIterator mnistTrain = new MnistDataSetIterator(batchSize, true,
randomSeed);
DataSetIterator mnistTest = new MnistDataSetIterator(batchSize, false,
randomSeed);
```

Model Configuration

The model we're going to use is a basic three-layer neural network with an input layer, a hidden layer, and then the output layer.

Our input layer is made up of 784 input nodes; this represents each pixel in the image (28 × 28), and the output of this input layer is sent to the hidden layer. The hidden layer is made up of 500 nodes and outputs to a 100-node output layer.

The final layer then uses the Softmax activation method to determine the output class, which consists of 10 different numbers in the MNIST dataset.

```
MultiLayerConfiguration conf = new NeuralNetConfiguration.Builder()
    .seed(randomSeed)
    .activation(Activation.RELU)
    .weightInit(WeightInit.XAVIER)
    .updater(new Nadam())
    .l2(rate * 0.005)
    .list()
    .layer(new DenseLayer.Builder()
            .nIn(imageHeight * imageWidth)
            .nOut(500)
            .build())
    .layer(new DenseLayer.Builder()
            .nIn(500)
            .nOut(100)
            .build())
    .layer(new OutputLayer.Builder(LossFunction.NEGATIVELOGLIKELIHOOD)
            .activation(Activation.SOFTMAX)
            .nOut(outputClasses)
            .build())
    .build();
```

With the configuration in place, it's assigned to the model and initialized. To see how the model is performing while training, we add a score iteration listening to the model. During the training, the score will be output to the console.

```
MultiLayerNetwork model = new MultiLayerNetwork(conf);
model.init();
model.setListeners(new ScoreIterationListener(5));
```

Model Training

To start the training of the model, we execute the `fit()` method with the training data set and the number of epochs we want to run.

```
model.fit(mnistTrain, numEpochs);
```

Once it's started, you will see the scores of the model update.

Model Evaluation

Once the training of the model is complete, it's time to evaluate the model with the test data from the MNIST dataset. The statistics from the evaluation will then be shown in the console.

```
Evaluation eval = model.evaluate(mnistTest);
log.info(eval.stats());
```

As you can see, the only real change from the neural networks created in Chapter 9 is how the input data is handled. The configuration, training, and evaluation are handled in the same way. Now that you've seen a basic multilayer perceptron configuration, let's cover a new algorithm type, convolutional neural networks, which are used widely in image processing applications.

Convolutional Neural Networks

The convolutional neural network (I'll refer to them as CNNs from this point on) was introduced into the machine learning word in 1998 where Yann LeCun, Leon Bottou, Yoshua Bengio, and Patrick Haffer published a paper outlining their work on a neural network called LeNet-5. As is tradition, the paper was using the MNIST number recognition dataset for both training and evaluation.

How CNNs Work

There are two parts to the CNN; the first is feature extraction, and the second is classification. Let's break these two elements down and examine them further.

Feature Extraction

The feature extraction element is the core part of the CNN algorithm. A convolution is a mathematical and integral function that will express the amount of overlap from one function to another.

These features could be elements of handwriting such as vertical or horizontal lines. If it were a car, it could be a wheel, a window, or a bumper. The convolution doesn't know what these things *actually* are; it's just noticing them in a numerical form.

Within a CNN the convolution is acting as a filter passing over the image. For example, if we were working with a 5 × 5–pixel image, such as Figure 11.4, the filter could be 3 × 3 pixels in size and scan the image from top left to bottom right, one pixel step across at a time.

1	0	1	1	0
0	1	1	1	1
0	0	0	1	1
1	1	1	0	0
0	1	0	1	0

Figure 11.4: 5 x 5–pixel image

The filter is a small matrix of values and layers over the image; the initial values are usually based on a random distribution (see Figure 11.5).

0	1	1
1	0	0
0	1	0

Figure 11.5: Filter matrix

By passing the filter over the image, the filter multiplies its own values with the values of the values on the input image. For example, if the filter was in the top-left position of the image, the filter calculation would look like this:

$$\left(1\text{x}0 + 0\text{x}1 + 1\text{x}1\right) + \left(0\text{x}1 + 1\text{x}0 + 1\text{x}0\right) + \left(0\text{x}0 + 0\text{x}1 + 0\text{x}0\right)$$
$$= 1$$

This is the receptive field—the output value of the CNN. The filter would move one step to the right and repeat the process until the entire image was covered. Note that the filter is overlapping the previous filter calculation; eventually you will end up with CNN output values for each step the filter has made (see Figure 11.6).

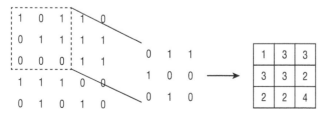

Figure 11.6: CNN output values

The filter values can be set in such a way that if we are looking for vertical features, then the vertical aspect of the filter could be set to 1, for example (see Figure 11.7). Or if looking for horizontal features, the filter could be set for that.

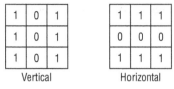

Figure 11.7: Filter values set at vertical and horizontal

How the filter is configured will determine the output of the CNN; features that are not featured in the filter will be dimmed from the outgoing computed CNN value.

Activation Functions

As we established in Chapter 9, there is an activation layer that is usually done by way of a Sigmoid function (or TANH if you prefer). The ANN is a linear combination of its own inputs.

Nonlinear functions let us perform classification of data even when it's not linearly separable. In the CNN, once the filter has passed the part of the image, it will be passed through to another mathematical function, an activation function.

The commonly used function for this operation is a Rectified Linear Unit, or ReLU for short. What the ReLU is doing is converting all the negative values to zero and leaving the positive values as they are. At this point, the resulting dimming created by the filter will be enhanced.

Pooling

The convolution layer, as described earlier, leaves us with a set of feature maps. Common among CNN implementations is to add a subsampling layer, also known as *pooling*. Performing pooling reduces the number of parameters and computations in the network and, therefore, reduces the dimensionality. With reduced dimension, this will decrease the time taken to train the network.

With max-pooling, we are looking at a window, similar to our feature filter, and moving across the feature map. This time, however, we are only finding the maximum value of the window and registering that value in the new filter.

Classification

So far, we've covered the steps of building the CNN; these are the convolution (our filter steps), using Rectified Linear Unit (ReLU) to enhance the feature map and then pooling to reduce the filter map again.

The CNNs are usually run over several iterations, so you end up with a process along the lines of: Convolution ⇨ ReLU ⇨ Max Pooling ⇨ Convolution ⇨ ReLU ⇨ Max-Pooling and then ending with some form of classification, to give you a fully connected layer.

The max-pooling outputs are fed into a neural network for classification. In most cases, this is a multilayer perceptron, the same that's covered in Chapter 9. The weights of the MLP will determine the classification.

The CNN is a powerful algorithm, and it's used widely for object detection, speech recognition, and image processing. There are various frameworks and architectures using CNNs, and it's worth taking the time to investigate how they are being implemented and used.

CNN Demonstration

Let's walk through a coded CNN. There are a number of steps to take in order to get the program working.

1. Acquire image data to train on.
2. Code the CNN.
3. Perform the training.
4. Save the generated model.

Downloading the Image Data

The Caltech101 image dataset is a training set comprised of 101 categories of images; there are between 40–800 images per category. For our use here, that's perfectly fine.

Download the .tar.gz file from

```
http://www.vision.caltech.edu/Image _ Datasets/Caltech101/
101_ObjectCategories.tar.gz
```

Unarchive the dataset to a location on your machine.

```
$ cp /your/downloads/folder/101_ObjectCategories.tar.gz /your/
data/folder
$ gunzip 101_ObjectCategories.tar.gz
$ tar xvf 101_ObjectCategories.tar
```

If you look at the directory, you will see the folders; these are going to be used at the label names for the model training.

```
$ ls -l
total 0
drwxr-xr-x@ 470 jasebell  staff  15040  9 Nov  2004 BACKGROUND_Google
drwxr-xr-x@ 437 jasebell  staff  13984  9 Nov  2004 Faces
drwxr-xr-x@ 437 jasebell  staff  13984  9 Nov  2004 Faces_easy
drwxr-xr-x@ 202 jasebell  staff   6464  9 Nov  2004 Leopards
```

```
drwxr-xr-x@  57 jasebell  staff   1824  9 Nov  2004 accordion
drwxr-xr-x@ 802 jasebell  staff  25664  9 Nov  2004 airplanes
drwxr-xr-x@  44 jasebell  staff   1408  9 Nov  2004 anchor
drwxr-xr-x@  44 jasebell  staff   1408  9 Nov  2004 ant
drwxr-xr-x@  49 jasebell  staff   1568  9 Nov  2004 barrel
drwxr-xr-x@  56 jasebell  staff   1792  9 Nov  2004 bass
.......
drwxr-xr-x@  41 jasebell  staff   1312  9 Nov  2004 wrench
drwxr-xr-x@  62 jasebell  staff   1984  9 Nov  2004 yin_yang
```

With the training data in place, we can now take a look at the code that will do the training. I'm going to break this down into step-by-step pieces and explain what's going on.

Basic Setup

The first step is to set up some constant values; these will be used in the training. The path to the training data, the image set you've downloaded and unarchived, is set.

```
public static final String pathToImages
"/path/to/data/101_ObjectCategories";
```

The image height, width, and color depth are set. As the images are RGB color, the depth is 3.

```
int imageHeight = 200;
int imageWidth = 300;
int channels = 3;
```

The next step is to set up the relevant values for the CNN itself; as with other neural nets, we set up a random seed value. The CNN will process in batches of 50 images. The output class value (which image category it is when predicted) is set to 101, which is the number of categories in the dataset.

```
int seed = 123;
Random randNumGen = new Random(seed);
int batchSize = 50;
int numOutputClasses = 101;
int epoch = 10;
```

Depending on your machine, you may get out-of-memory errors during training; this isn't because of the Java Virtual Machine but another dependency used by Apache OpenCV. To avoid the program from halting during training, I've found setting the properties of the following two values solves the problem:

```
System.setProperty("org.bytedeco.javacpp.maxphysicalbytes", "0");
System.setProperty("org.bytedeco.javacpp.maxbytes", "0");
```

Handling the Training and Test Data

The model needs to know the file paths for all the files. The `FileSplit` class takes all the file paths and makes sure they are suitable for training. We're looking for images.

```
File trainingData = new File(pathToImages);
FileSplit trainingDataSplit = new FileSplit(trainingData,
NativeImageLoader.ALLOWED_FORMATS, randNumGen);
```

The folder names within the dataset directory are going to be used as the names for the output classes. The `ParentPathLabelGenerator` handles the directory names and converts them to class names for the model.

```
ParentPathLabelGenerator labelMaker = new ParentPathLabelGenerator();
```

We want the CNN to be evenly balanced across the class label directories; some may contain more images than others, for example. For a good balanced model, we need to ensure the training data is balanced too. This is handled by the `BalancedPathFilter` class. After that is set, we then split the input data into a training data set with 80 percent of the image data and a test data set with the remaining 20 percent.

```
BalancedPathFilter pathFilter = new BalancedPathFilter(randNumGen,
labelMaker,0, numOutputClasses, batchSize);
InputSplit[] filesInDirSplit = trainingDataSplit.sample(
pathFilter, 80, 20);
InputSplit trainingSet = filesInDirSplit[0];
InputSplit testingSet = filesInDirSplit[1];
```

Image Preparation

We set up three image transformations that will be used in the training. In the first two, the image is flipped either by its x- or y-axis, or not at all. The last transform of the image is randomly warped.

```
ImageTransform tf1 = new FlipImageTransform(randNumGen);
ImageTransform tf2 = new FlipImageTransform(new Random(seed));
ImageTransform warptransform = new WarpImageTransform(randNumGen,42);
List<ImageTransform> tranforms = Arrays.asList(new ImageTransform[] {
tf1, warptransform, tf2});
```

The images are loaded by the `ImageRecordReader`, passing in the height, the width, the number of channels, and the previously created label maker. Once initialized, the program now has the information for that image set.

```
ImageRecordReader recordReader = new ImageRecordReader(imageHeight,
imageWidth, channels, labelMaker);
recordReader.initialize(trainingSet);
```

Lastly, we reduce the scale from the values 0 to 255, down to 0 to 1.

```
DataNormalization scaler = new ImagePreProcessingScaler(0, 1);
    DataSetIterator dataSetIterator;
```

CNN Model Configuration

Now it's time to create the CNN model configuration, shown in Listing 11.1. There are 14 layers to this neural network. See Table 11.2 for the layer steps.

Table 11.2: CNN Layers

LAYER	FUNCTION
1	Convolution layer, with a 1 × 1 pixel filter step with 3 input nodes and 16 output nodes.
2	Normalization layer.
3	ReLU Activation layer.
4	Convolutional layer with another 1 × 1-pixel filter step, this time with 16 input nodes and 16 output nodes.
5	Normalization layer.
6	ReLU Activation layer.
7	Pooling layer.
8	Convolution layer with 2 × 2-pixel filter step, 16 input nodes, and 16 output nodes.
9	Normalization layer.
10	ReLU Activation layer.
11	Pooling layer.
12	A drop-out layer; features that are not important to the CNN are dropped.
13	The Dense layer flattens the data and combines with all the neurons.
14	The Softmax layer then gives the classifier on all 101 output classes.

Listing 11.1: CNN Model Configuration

```
MultiLayerConfiguration conf = new NeuralNetConfiguration.Builder().
trainingWorkspaceMode(WorkspaceMode.SEPARATE)
        .inferenceWorkspaceMode(WorkspaceMode.SINGLE)
        .seed(seed)
        .iterations(1)
        .activation(Activation.IDENTITY).weightInit(WeightInit.XAVIER)
        .optimizationAlgo(OptimizationAlgorithm.STOCHASTIC_GRADI
```

```
ENT_DESCENT)
        .learningRate(.006)
        .updater(Updater.NESTEROVS)
        .regularization(true).l2(.0001)
        .convolutionMode(ConvolutionMode.Same).list()
        // block 1
        .layer(0, new ConvolutionLayer.Builder(new int[] {5, 5})
.name("image_array").stride(new int[]{1, 1})
                .nIn(3)
                .nOut(16).build())
        .layer(1, new BatchNormalization.Builder().build())
        .layer(2, new ActivationLayer.Builder(
).activation(Activation.RELU).build())
        .layer(3, new ConvolutionLayer.Builder(new int[] {5, 5})
.stride(new int[]{1, 1}).nIn(16).nOut(16)
                .build())
        .layer(4, new BatchNormalization.Builder().build())
        .layer(5, new ActivationLayer.Builder()
.activation(Activation.RELU).build())
        .layer(6, new SubsamplingLayer
.Builder(SubsamplingLayer.PoolingType.AVG,
                new int[] {2, 2}).build())
        .layer(7, new ConvolutionLayer.Builder(new int[] {5, 5})
.stride(new int[]{2, 2}).nIn(16).nOut(16)
                .build())
        .layer(8, new BatchNormalization.Builder().build())
        .layer(9, new ActivationLayer.Builder()
.activation(Activation.RELU).build())
        .layer(10, new SubsamplingLayer
.Builder(SubsamplingLayer.PoolingType.AVG,
                new int[] {2, 2}).build())
        .layer(11, new DropoutLayer.Builder(0.5).build())
        .layer(12, new DenseLayer.Builder().name("ffn2").nOut(256)
.build())
        .layer(13, new OutputLayer
.Builder(LossFunctions.LossFunction.NEGATIVELOGLIKELIHOOD)
                .name("output").nOut(numOutputClasses)
.activation(Activation.SOFTMAX).build())

        .setInputType(InputType.convolutional(imageHeight, imageWidth,
channels))
        .backprop(true)
        .pretrain(false)
        .build();
```

Each step on the CNN is a layer. With the configuration done, it's time to move on and use this configuration to create the model.

```
MultiLayerNetwork model = new MultiLayerNetwork(conf);
model.init();
```

During the training, we want to see some progress of how the model is doing. The `ScoreIterationListen` will give us the output; this is set to the model.

```
model.setListeners(new ScoreIterationListener(10));
 //To see our model's progress
```

With the model initialized, we can now start training.

Model Training

The training happens on each of the image transformations that were defined earlier: two flipped image transformations and the warped image transformation.

```
for (ImageTransform transform: tranforms) {
    System.out.println("Training using image transform: "
+transform.getClass().toString());
    recordReader.initialize(trainingSet,transform);
    dataSetIterator = new RecordReaderDataSetIterator(recordReader,
batchSize, 1, numOutputClasses);
    scaler.fit(dataSetIterator);
    dataSetIterator.setPreProcessor(scaler);
    for (int j = 0; j < epoch; j++) {
        model.fit(dataSetIterator);
    }
}
recordReader.reset();
```

With that complete, the training is then performed on the original image dataset that was registered with the training set.

```
recordReader.initialize(trainingSet);
dataSetIterator = new RecordReaderDataSetIterator(recordReader,
batchSize, 1, numOutputClasses);
scaler.fit(dataSetIterator);
dataSetIterator.setPreProcessor(scaler);
for (int j = 0; j < epoch; j++) {
    model.fit(dataSetIterator);
}
```

Please note this training will take some time to complete. Depending on the specification of your machine, it's probably a good time to make several cups of tea or coffee, tidy up the house, or even go for a walk.

Model Evaluation

With the model trained, it's now time to evaluate the model with the other 20 percent of the dataset. The record reader is reset and then initialized with the training dataset.

The evaluation is then created and run against the test dataset. Results of the evaluation are then output to the console.

```
recordReader.reset();
recordReader.initialize(testingSet);
DataSetIterator testIter = new RecordReaderDataSetIterator(recordReader,
batchSize,1,numOutputClasses);
scaler.fit(testIter);
testIter.setPreProcessor(scaler);

Evaluation eval = new Evaluation(numOutputClasses);
while(testIter.hasNext()){
    DataSet next = testIter.next();
    INDArray output = model.output(next.getFeatureMatrix());
    eval.eval(next.getLabels(),output);

}

log.info(eval.stats());
```

Saving the Model

The final step of our CNN is to persist the model to a zip file so it can be reused for other predictions.

```
File locationToSave = new File(System.getProperty("user.dir"),
"mycnntest.zip");
ModelSerializer.writeModel(model,locationToSave,false);
```

Transfer Learning

So far, we've covered two potential ways of classifying images through neural networks: the multilayer perceptron and convolutional neural network. As you may have noticed, these can take time to train. Using a technique called *transfer learning*, we can use existing models and use their weights to do predictions or create a newly updated model based on the pretrained model.

Model training on images and video are intensive on your time and your computing power, so finding that someone else has done all the work so we can reuse the model is great. Keep in mind that if you are training ImageNets with a large number of images, the training time can take weeks.

The DeepLearning4J website has a model zoo that lets you use existing models in your work. You'll see examples for CNNs that you've just learned about and other model types such as Long Short Term Memory (LSTM) models.

Convolutional neural networks do not show the specific features until the later layers; this means that most of the generic training happens up front. There are two possible processes we can use to make use of what's already out there.

With something like a CNN, we can make use of the classifier layer and replace it with the version of the classifier we want to use. This means we are not replicating huge amounts of training where the heavy work is happening, meaning the convolution, normalization, pooling, and activation steps.

Within DeepLearning4J, there is a TransferLearning API that will help you refine the output classifier and save the newly trained model. This is useful for ImageNet-like models.

If you are in a situation where you find yourself with a small training dataset or a dataset that you know is similar to features that were present in the originally trained model, then applying transfer learning will save you development and training time. Where possible, always reuse.

Summary

In this chapter, we covered using neural networks to process images. First we used the multilayer perceptron to work on the MNIST number data set and then we introduced the more complete convolutional neural network for a more thorough way of extracting features from image data.

While these approaches are good to know in theory, their training and application times can run into hours and days, depending on the volume of images and the way you want to train your model. So, if possible, seek out pretrained models and investigate using transfer learning to update the classifier output layer to match the domain you're working in.

In Chapter 12 we bring in a streaming data application, Kafka, and investigate how to develop a self-training and updating machine learning system.

Machine Learning Streaming with Kafka

Within this book we have mainly concentrated on machine learning as a singular process: acquiring data, training a model, and then making predictions. In this chapter, we will look at streaming data and how that can affect model training and predictions.

What You Will Learn in This Chapter

This is an involved chapter with a lot of code examples and process to work through. By the end of the chapter, you'll have a proof-of-concept application that takes in streaming data, events, and multiple machine learning models.

First, we'll cover setting up Kafka, one of the main applications used in streaming log-like data. Then we'll move on to how topics work and the methods used to produce and consume messages. Lastly, we'll look at how these fit into a machine learning framework and design a system that continuously trains itself and handles multiple prediction models.

This chapter does not concentrate on how the models work but on how they can be involved in an actual application. The perspective changes from machine learning to engineering. It will also bridge several JVM-based languages like Java and Clojure and present a path of least resistance to easily create REST-based APIs that can be called from Kafka-based applications.

From Machine Learning to Machine Learning Engineer

It's fair to say that some of the concepts covered in this book have been around for a long time. Neural networks have been around for more than 50 years, for example. What has changed over that time period is the growth of computing power and other related factors such as the reduced cost of storage. It's become easier to process more data cheaply over time. This brings with it some interesting developments in how quickly we can process data, run algorithmic computations, and, in turn, generate predictions to stakeholders.

These types of system, however, are not easy to build. Cloud-based infrastructures such Amazon's AWS Sagemaker, Google's ML Engine (now Cloud AI), and Microsoft's Azure Machine Learning platform aim to make things easier, but the skills required to get any of these solutions performing well are still hard to come by.

The rise of the data engineer or machine learning engineer as a job function is becoming more important than ever before. A hybrid of a software developer and a system administrator and a smattering of a data scientist, the machine learning engineer now stands between a machine learning model being created and how it's deployed into production-based systems.

Engineering skills are always in demand, and anyone who can acquire, clean, and prepare data for production and piece together the required components to create an end-to-end solution is like gold dust on a team. You'll see all manner of blog posts telling you what you should know to be an ML engineer, but the basics are simple when they are broken down.

- Knowledge in how machine learning models work
- Able to prepare data
- Proficient in at least one development language, preferably the one that those creating the models are using
- Core Linux skills (do you know your `cat` from your `ls`?)
- Deployment methods and being able to install all manner of tools to get this end-to-end solution working

There will always be raging debates on whether it's better to be a specialist or a generalist. Personally, I've found it more helpful to know as much as I can, so I've always considered myself a generalist. I'd never even considered being a machine learning engineer until I was told by a manager that's basically what I was doing.

From Batch Processing to Streaming Data Processing

When the first edition of this book was written, the explosion of Big Data technologies was all over the technical publications, blogs, and news sites. From the benefits to the scary stories (predicting pregnancy from shopping cart data, for example), you could not escape that Hadoop was changing the world.

The emphasis was on batch data. Huge piles of structured and unstructured data could be stored on Hadoop's filesystem (HDFS) or a storage solution like Amazon Web Services Simple Storage System (S3).

When Spark arrived, it made big leaps in improving the performance in the Big Data systems. In addition, it provided an effective SQL-like way of querying the data with SparkQL and a streaming system called Spark Streaming. The streaming system wasn't exactly what we'd considered true streaming; it was based on windows of batch data processed by the Spark job. This is best described as *micro-batching*. It works well and is still widely used.

The increase in the volume and velocity of Internet of Things (IoT) data and social media data has increased the need for true streaming frameworks to handle the data. A number of streaming frameworks existed including RabbitMQ, Storm, and SpringXD (which I covered in the first edition of this book but reached end of life in 2017), plus more traditional message brokers like ActiveMQ. One name that kept being repeated was Apache Kafka, and that's what we'll concentrate on in this chapter.

What Is Kafka?

Kafka is a stream processing platform. It was originally developed by LinkedIn and then open sourced in 2011. Kafka provides a high-performance, fault-tolerant streaming data service. It acts on a publisher/subscriber (pub/sub) message queue, and if you want real-time data feeds, then Kafka is an option that should be seriously considered.

How Does It Work?

The true power behind Kafka is that it's scalable and can be run on one or multiple servers, known as *brokers*. Messages are sent to topics, producers send messages to the broker, and consumers take messages from the broker.

To the producers and consumers subscribed to the system, it would appear as a stand-alone processing engine, but production systems are built on many machines. It can handle millions of messages of throughput, dependent on the physical disk and RAM on the machines, and is fault tolerant.

Messages are sent to topics, which are written sequentially in an immutable log. Kafka can support many topics and can be replicated and partitioned (see Figure 12.1).

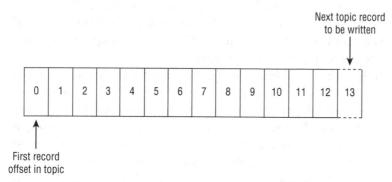

Figure 12.1: Topics written sequentially in Kafka

Once the records are appended to the topic log, they can't be deleted or amended. It's a simple data structure where each message is byte encoded. Producers and consumers can serialize and deserialize data to various formats.

Messages are sent to the Kafka cluster by producers, and these messages are stored by the broker in topics. Consumers subscribe to topics and poll the topic for messages to read. The broker nodes are dumb; it's the producers and consumers that are doing the smart work (see Figure 12.2).

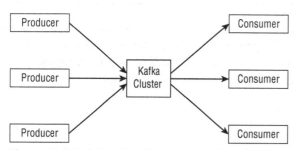

Figure 12.2: Relationship of producers to the Kafka cluster and consumers

Topics can grow in size, and they can be split into partitions (see Figure 12.3). As far as the producers and consumers are concerned, it's a single log of messages, and there's no concern about which partition the messages are in. The important thing to remember is that messages are read from the consumer in the order they were sent from the producer.

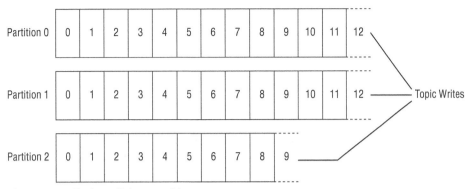

Figure 12.3: Topics split into partitions

Fault Tolerance

Partition data is replicated over a number of brokers. If a broker dies, then data is still available over the remaining brokers. One broker is designated the leader and is the node that all data to the partition is written to and read from.

If the leader fails, then the cluster will assign a new lead broker, and the other brokers will follow that one. The registry of brokers, including which is the leader, is maintained by Zookeeper.

Further Reading

For more information about the workings of Kafka, I've included some reading material in Appendix D, "Further Reading." In this chapter, I concentrate on the more practical aspects of getting Kafka running and then doing machine learning with it.

Installing Kafka

There are two versions of Kafka that can be downloaded. There's the open source Apache version, and there's the Confluent Community version. They both operate in the same way; the main difference is when we bring the SQL query language KSQL into the mix.

For this chapter, I'm going to use the Apache Kafka distribution, which is available at `https://kafka.apache.org`. This is the download link that I'm using:

```
https://www.apache.org/dyn/closer.cgi?path=/kafka/2.2.0/kafka _ 2.12-
2.2.0.tgz
```

You will need to choose a mirror website to download your Kafka distribution from.

Download the TGZ file and store it in the directory where you want to install Kafka. Installing Kafka is simple; the first step is to expand the zipped TAR file into a directory.

```
$ tar -zxf kafka_2.12-2.2.0.tgz
$ cd kafka_2.12-2.2.0
```

As Kafka relies on Zookeeper for broker identification, Zookeeper needs to be running *before* we start Kafka. There are various configuration files that are used within Kafka, but there are two specifically that I'll focus on here: `zookeeper.properties` and `server.properties`. How these files are used depends on the type of cluster you want to operate. I will cover two in this book: a single-node cluster and a multinode cluster running on the same machine.

Kafka as a Single-Node Cluster

It's always a good idea to practice on a single-node cluster configuration of Kafka. A single machine will handle both Zookeeper and Kafka. I use the single-node configuration for developing and testing producers, consumers, and streaming applications. Once I'm happy that the applications are performing as expected, then I can look at deploying them on a multinode cluster.

The first application to start is Zookeeper. Kafka requires Zookeeper to maintain the state of the brokers. If it's not live, then Kafka will not work. In development, I have multiple terminal window sessions open—one for Zookeeper, one for Kafka, and another for whatever producer or consumers I'm working on.

Starting Zookeeper

Let's start a single-node Kafka installation. The first thing to do is open a terminal window and go to the directory where you installed Kafka. From the command line, type **bin/zookeeper-server-start.sh config/zookeeper.properties**.

```
$ bin/zookeeper-server-start.sh config/zookeeper.properties
[2019-06-18 07:43:12,238] INFO Reading configuration from: config/
zookeeper.properties (org.apache.zookeeper.server.quorum.
QuorumPeerConfig)
[2019-06-18 07:43:12,241] INFO autopurge.snapRetainCount set to 3 (org.
apache.zookeeper.server.DatadirCleanupManager)
[2019-06-18 07:43:12,241] INFO autopurge.purgeInterval set to 0 (org.
apache.zookeeper.server.DatadirCleanupManager)
[2019-06-18 07:43:12,243] INFO Purge task is not scheduled. (org.apache.
zookeeper.server.DatadirCleanupManager)
[2019-06-18 07:43:12,243] WARN Either no config or no quorum defined in
config, running  in standalone mode (org.apache.zookeeper.server.quorum.
QuorumPeerMain)
```

```
[2019-06-18 07:43:12,260] INFO Reading configuration from: config/
zookeeper.properties (org.apache.zookeeper.server.quorum.
QuorumPeerConfig)
```

Starting Kafka

Once Zookeeper is running, you can now start the Kafka broker. Open a new terminal window and go to the same directory from where you started Zookeeper. Now type **bin/kafka-server-start.sh config/server.properties**.

```
$ bin/kafka-server-start.sh config/server.properties
[2019-06-18 07:43:45,792] INFO Registered kafka:type=kafka.
Log4jController MBean (kafka.utils.Log4jControllerRegistration$)
[2019-06-18 07:43:46,536] INFO starting (kafka.server.KafkaServer)
[2019-06-18 07:43:46,538] INFO Connecting to zookeeper on localhost:2181
(kafka.server.KafkaServer)
[2019-06-18 07:43:46,562] INFO [ZooKeeperClient] Initializing a new
session to localhost:2181. (kafka.zookeeper.ZooKeeperClient)
[2019-06-18 07:43:46,568] INFO Client environment:zookeeper.
version=3.4.13-2d71af4dbe22557fda74f9a9b4309b15a7487f03, built on
06/29/2018 00:39 GMT (org.apache.zookeeper.ZooKeeper)
[2019-06-18 07:43:46,568] INFO Client environment:host.
name=192.168.1.102 (org.apache.zookeeper.ZooKeeper)
[2019-06-18 07:43:46,568] INFO Client environment:java.version=1.8.0_45
(org.apache.zookeeper.ZooKeeper)
```

Kafka is now ready for use, although it's not doing anything useful at the moment; we'll move on to that shortly. In this development setup, the correct routine to close it down is to close Kafka first (Ctrl+C will stop the application) and then do the same with Zookeeper. If you close Zookeeper while Kafka is running, the broker will not be able to connect to anything and start feeding out errors in the console.

Kafka as a Multinode Cluster

Whereas the single-node cluster has just one broker connected to Zookeeper, a multinode cluster has many brokers connected to Zookeeper, and the log is distributed across the cluster. You can create this cluster locally, but it's not a configuration that should really be used in a production setting. It does, however, give a good foundation for how the multinode cluster works. In this example, I'll set up a three-broker configuration running on a single Zookeeper cluster.

There are some configuration aspects that we need to do before we start Kafka. In the config directory, there is the server.properties file; this is the configuration for a single broker. If you open the file and look for the broker.id key, you will see the value 0 for it. The comments also give important information.

```
############################## Server Basics ##############################

# The id of the broker. This must be set to a unique integer for each broker.
broker.id=0
```

The keyword here is *unique*, so we need another two broker IDs (1 and 2 for this example). Our first task is to create a copy of the server properties file for the other two brokers.

```
$ cp server.properties server1.properties
$ cp server.properties server2.properties
```

Each of these new configuration files needs editing. Open your text editor, edit the `server1.properties` file so that it has the same values as in Table 12.1, and then make the changes in `server2.properties`.

Table 12.1: Server 1 and 2 Properties

SETTING NAME	SERVER1.PROPERTIES	SERVER2.PROPERTIES
broker.id	1	2
listeners	PLAINTEXT://:9093*	PLAINTEXT://:9094*
log.dirs	/tmp/kafka-logs1	/tmp/kafka-logs2

* Remove the comment character (#) from the listeners line of the properties file.

What have we changed? There are three changes. First, there's the broker ID as these must be unique, so we now have broker 0, 1, and 2. Second, each broker needs a port address that's different from the other brokers as they are going to be operating on the same machine. Last, the directory of the log information, the actual messages, changed, as each broker needs its own location for the messages and other associated files.

Starting the Multibroker Cluster

As with the stand-alone node, Zookeeper needs starting. Open a terminal window and run the startup command as before:

```
$ bin/zookeeper-server-start.sh config/zookeeper.properties
```

To start the brokers, open three new terminal windows and ensure that each is in the directory where Kafka was installed. One broker is started in each terminal window.

Starting the first broker, broker 0, is the same as previously shown.

```
$ bin/kafka-server-start.sh config/server.properties
```

The remaining brokers can now be started; next is broker 1.

```
$ bin/kafka-server-start.sh config/server.properties1
```

Then comes broker 2 (see Figure 12.4).

```
$ bin/kafka-server-start.sh config/server.properties2
```

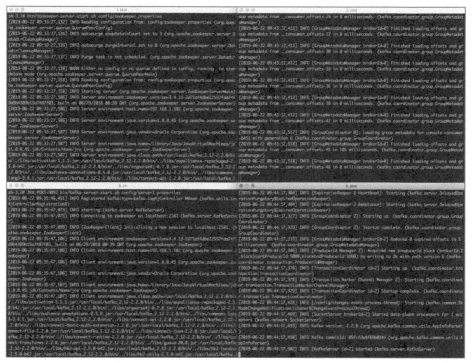

Figure 12.4: Multibroker cluster

For the remainder of this chapter any commands or programs run using Kafka will assume that your Kafka broker is running in stand-alone mode and using localhost as the hostname.

Topics Management

Any message that you send within the Kafka system is sent to a topic. You are free to create as many topics as you want. There are some general rules of thumb about the optimum number of topics you can have on a cluster depending on how many nodes you have.

Within the `bin` directory of your Kafka distribution, there are several shell scripts that will do the required tasks for topic management. The main one to focus on here is the `kafka-topic.sh` command.

In this short introduction, you will see how to create, list, and delete topics and how to send messages to the log and receive them.

Creating Topics

Topics can be created in one of two ways. The first is with the command-line tool, and the second is from within a producer application, which will create the topic if it does not exist.

It's worth working with the command line as that will give you an understanding of what's required within the topic creation, the required settings for replication and partitions, and how the broker and Zookeeper settings are handled.

Let's create a test topic now; we'll call it `testtopic` for ease of demonstration. Within the `bin` directory of the Kafka distribution, you will see a shell script file called `kafka-topic.sh`. This is the main command for topic-related activities. To create the topic, execute the following command from your terminal window:

```
$KAFKA_HOME/bin/kafka-topics.sh --zookeeper localhost:2181 --create
--topic testtopic --replication-factor 1 --partitions 1
Created topic "testtopic".
```

Let's break this command down into smaller segments and break down what the script is doing. The first setting is telling Kafka which Zookeeper server we are using. This is a required setting, as we're using the local deployment, and the Zookeeper server is running on localhost on port 2181 (the default port for Zookeeper).

Next `--create` is telling the application we want to create a topic. The name of the topic is handled next with the `--topic` flag. As the cluster is just for development, I'm not concerned about the replication factor or the number of partitions, so this is set to 1.

Finding Out Information About Existing Topics

With the same command, you can find out information about the topics that are available on the cluster. To generate a list of all the topics on the cluster, run the following command:

```
bin/kafka-topics --zookeeper localhost:2181 --list
testtopic
```

If you've created only one topic, then you may be surprised to see more topics than you were expecting. The Kafka cluster creates internal topics for monitoring and metrics. All the offset information, including the message positions of the associated consumers, is held within a topic as well.

The application can expand on the information for the topic by using the `--describe` flag.

```
$ bin/kafka-topics --zookeeper localhost:2181 --describe --topic testtopic
Topic:testtopic     PartitionCount:1     ReplicationFactor:1     Configs:
    Topic: testtopic    Partition: 0    Leader: 0    Replicas: 0    Isr:
0
```

Describing the topic produces a multiline output that expands on how Kafka is using the cluster for that topic. On the first line is the partition count and replication factor. In this example it's only 1, as it was when the topic was created earlier.

The second line gives more information on the broker ID that's the leader and the IDs of the broker replicas.

Deleting Topics

Topics, regardless of whether they have messages in them or not, use resources on the cluster. It's good practice to remove topics that you are not using. The deletion of topics is done through the `kafka-topics` command-line program, but will work only if `delete.topic.enable` is set to true.

Any other value for that setting will mean that the delete topic command will be ignored.

```
$ bin/kafka-topics --zookeeper localhost:2181 --delete --topic testtopic
Topic testtopic is marked for deletion.
Note: This will have no impact if delete.topic.enable is not set to true.
```

Sending Messages from the Command Line

The best way to test that the cluster is working is by using the command-line tools to create and consume messages. The `kafka-console-producer` application enables you to send messages from the command line.

```
$ bin/kafka-console-producer --broker-list localhost:9092 --topic testtopic
```

Once the application starts, you can start typing data into the terminal window. Each time you press the Return key, your message is sent to the broker with the desired topic. If you want to exit the application, press the key combination of Ctrl+C, and the application will stop and return you to the terminal command line.

With some test data in Kafka, now it's time to turn our attention to reading the messages in the log.

Receiving Messages from the Command Line

As well as a command-line application to create messages and push them to the message log, there is the equivalent to read messages from the log too.

```
$ bin/kafka-console-consumer --bootstrap-server localhost:9092 --topic
testtopic --from-beginning
```

The `--from-beginning` flag reads messages from the start of the message log. If you leave it out, the application will read messages from the last offset position. The application will continue running until you close it with Ctrl+C.

Kafka Tool UI

For those who aren't used to or just don't like using the command line, there are tools available that will make looking after the Kafka cluster a little easier.

If you are using the Confluent Kafka distribution, you have access to the control center, which is a web-based tool for visualizing your cluster (see Figure 12.5). To access it, point your web browser to the IP address of your cluster along with port 9021, for example.

```
http://192.168.1.102:9021
```

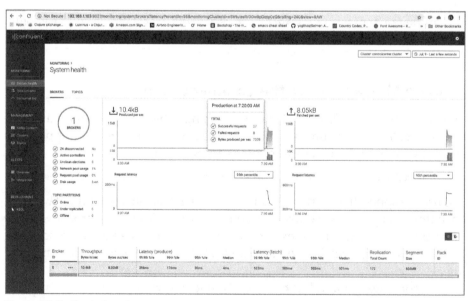

Figure 12.5: Control center

As I can never tell what type of Kafka cluster I'll be working on, I use a generic tool called Kafka Tool (see Figure 12.6), which is available from `http://www.kafkatool.com` and free for personal use. It runs on macOS, Linux, and Windows.

Once it's installed, you have an overview of your cluster and can inspect the general cluster information, get the Zookeeper status, and view your topic and consumer group information. It's useful for inspecting message contents and checking the offset information of topics too. Sometimes this is hard to read from the command line, and a tool such as this one can make life a little easier.

Figure 12.6: Kafka Tool

Writing Your Own Producers and Consumers

So far, the emphasis has been on the Kafka brokers and the command-line tools to do some basic tasks such as create topics and manage them. Now it's time to concentrate on writing some basic applications to send messages to the brokers and read messages from them too.

Various client libraries cover the majority of languages; for a list of what's available, take a look at the list here:

```
https://cwiki.apache.org/confluence/display/KAFKA/Clients
```

For this chapter, there are examples of producers and consumers in Java. Later in the chapter, you'll see Kafka code written in both Java and Clojure; they both run off the same client APIs.

Producers in Java

As you'll see in both these sections, the process of producing and consuming messages is quite simple. There is some preliminary setup that is required before you can start working with the broker. Each segment of the application is explained first, and you can then see the completed code.

Properties

Any Java application that communicates with a Kafka cluster is set up via the Java `Properties` class. The key-pair values are required by the Kafka API to send default settings and flags to the brokers.

```
Properties props = new Properties();
    props.put("bootstrap.servers", "localhost:9092");
    props.put("acks", "all");
    props.put("retries", 0);
    props.put("key.serializer",
        "org.apache.kafka.common.serialization.StringSerializer");
    props.put("value.serializer",
        "org.apache.kafka.common.serialization.StringSerializer");
```

Looking at this properties setup, we can see some settings that are familiar from previous examples in the chapter but also some new ones.

Bootstrap Servers

The `bootstrap.servers` key is telling the application where the brokers are. In the stand-alone development cluster, this is only one node and hence one bootstrap server. It acts as a starting point for the application to discover the other brokers in the cluster. If you have more than one broker, you can use comma-separated servers in the value.

```
props.put("bootstrap.servers", "server1.localhost:9092, server2.localhost:
9093");
```

Acks

Next is the key `acks`, which is short for acknowledgments. When a producer sends a message to the broker, it can receive an ack to say that the broker has received the message. While this is good from an application safety point of view, it's worth keeping in mind there is a small amount of latency introduced. Three settings are available for the `acks` key. Table 12.2 explains the settings.

Table 12.2: Kafka Producer Acks

ACKS SETTING	MEANING
acks=0	The producer does not receive any confirmation that the broker has received the message. This is basically a fire-and-forget setting, and there is no guarantee the message has reached the broker.
acks=1	The producer will receive an acknowledgment from the lead broker (the leader), but it will have been received without the confirmation that all the brokers that follow that leader have received the message.
acks=all or acks=-1	The producer will receive an acknowledgment once the leader has received confirmation that the message is with the in-sync replicas.

Retries

The producer client will try to resend the message when a failed send occurs. In this example, the zero value means the client isn't going to attempt a resend. Using retries can cause issues with message ordering. If a client fails at sending a batch of messages and then succeeds with a second batch, there is a chance that the second message batch will be available before the first.

Deserializers

Kafka treats messages as a binary payload. This means that messages need to be serialized by a producer and then desterilized by the consumer when the message is eventually read. The properties configuration sets the producer serializer, which is in two parts; there's a key and a value. In this example, there's a string-based key, and the message payload is a string too.

The Producer

With the configuration all set up in the `Properties` class, it's now time to create the actual producer. The `Producer` class takes the key and value object types that are specified in the key and value serializer settings. The `Producer` is created in one line, passing in the configuration properties.

```
Producer<String, String> producer = new KafkaProducer
        <String, String>(props);
```

Messages

Now we're at the stage where we can send messages to the broker. The `send()` method sends a `ProducerRecord` object, which takes the topic name, the key, and the message.

```
producer.send(new ProducerRecord<String, String>(topicName,
        keyName, payLoad));
            System.out.println(keyName + " - Message sent
successfully: " + payLoad);
            producer.close();
```

The Final Code

Those three steps—configuring the properties, creating a producer, and then sending the message—create the basic components of a producer application. The following code is the finished application; notice that there is a `for` loop in there to send 10 messages to the broker.

```java
import java.util.Properties;
import org.apache.kafka.clients.producer.Producer;
import org.apache.kafka.clients.producer.KafkaProducer;
import org.apache.kafka.clients.producer.ProducerRecord;

public class ExampleProducer {
    public static void main(String[] args) throws Exception{
        String topicName = "testtopic";

        Properties props = new Properties();
        props.put("bootstrap.servers", "localhost:9092");
        props.put("acks", "all");
        props.put("retries", 0);
        props.put("batch.size", 16384);
        props.put("linger.ms", 1);
        props.put("buffer.memory", 33554432);
        props.put("key.serializer",
            "org.apache.kafka.common.serializa-tion.StringSerializer");
        props.put("value.serializer",
            "org.apache.kafka.common.serializa-tion.StringSerializer");

        Producer<String, String> producer = new KafkaProducer
            <String, String>(props);

        for(int i = 0; i < 10; i++)
            producer.send(new ProducerRecord<String, String>(topicName,
                Integer.toString(i), Integer.toString(i)));
                System.out.println("Message sent successfully");
                producer.close();
    }
}
```

Message Acknowledgments

The responsibility for processing the acks comes down to the producer application. Using the `ProducerInterceptor` class in Java, for example, gives you a callback method that is triggered when the ack is acknowledged.

```java
import org.apache.kafka.clients.producer.ProducerInterceptor;
import org.apache.kafka.clients.producer.ProducerRecord;
import org.apache.kafka.clients.producer.RecordMetadata;
import org.slf4j.Logger;
import org.slf4j.LoggerFactory;
import java.util.Map;

public class ProducerWithCallback implements ProducerInterceptor{
    private int onSendCount;
    private int onAckCount;
```

```
      private final Logger logger = LoggerFactory.
getLogger(ProducerWithCallback.class);

    @Override
    public ProducerRecord onSend(final ProducerRecord record) {
        onSendCount++;

        System.out.println(String.format("onSend topic=%s key=%s value=%s
%d \n",
                record.topic(), record.key(), record.value().toString(),
                record.partition()
        ));
        return record;
    }

    @Override
    public void onAcknowledgement(final RecordMetadata metadata, final
Exception exception) {
        onAckCount++;
        System.out.println(String.format("onAck topic=%s, part=%d, offset=
%d\n",
                metadata.topic(), metadata.partition(), metadata.offset()
        ));
    }

    @Override
    public void close() {
        System.out.println("Total sent: " + onSendCount);
        System.out.println("Total acks: " + onAckCount);
    }

    @Override
    public void configure(Map<String,?> configs) {
    }
}
```

Consumers in Java

As with producers, the consumer code is straightforward. Let's break down
the steps again.

Properties

The properties, as before, set up the various configuration settings for the
consumer application. The application needs to know the Kafka cluster to connect to; this is done with the `bootstrap.server` setting.

Consumer applications are members of consumer groups. It's not essential to specify one, but Kafka will assign one during runtime if a group name is not given. It's worth using your own group names for ease of searching for metrics on consumers.

As previously shown in the producer application, we are aware that the key and payload data for our messages are both strings. The producer application serialized the data with the `key.serializer` and `value.serializer` settings. Consumers require the deserializers for the message. When the message is consumed by the consumer, it arrives as a byte array; it's then deserialized to the format specified in the properties.

```
Properties props = new Properties();

    props.put("bootstrap.servers", "localhost:9092");
    props.put("group.id", "testcg");
    props.put("enable.auto.commit", "true");
    props.put("auto.commit.interval.ms", "1000");
    props.put("session.timeout.ms", "30000");
    props.put("key.deserializer",
        "org.apache.kafka.common.serialization.StringDeserializer");
    props.put("value.deserializer",
        "org.apache.kafka.common.serialization.StringDeserializer");
```

Fetching Consumer Records

Consumer applications poll records from the broker. The `.poll()` method takes an integer value with the number of records you want the consumer to process. Depending on the message load from the brokers, if you pull too many messages, you can have a slow-running consumer. Take time to test and experiment with the number of records you are polling.

During the polling and processing of records, the commit position of the topic is being written back to the log. If the properties of the consumer enable the auto setting of the message offset, this will be performed while your application is running. The duration between each setting of the offset is determined by the value of the `auto.commit.interval.ms` setting; in this application example, it's set to 1000ms, or 1 second.

```
KafkaConsumer<String, String> consumer = new KafkaConsumer
        <String, String>(props);
ConsumerRecords<String, String> records = consumer.poll(100);
```

The Consumer Record

With the `ConsumerRecords` class polling the broker and fetching records to process, the last step is to do something with the message contents. The `ConsumerRecord` class is the representation of a single record.

Each consumer record contains the following information:

- The topic name
- The partition number of where the record was received
- The offset position in the partition
- The timestamp as created by the producer
- The key value
- The actual message payload

You don't need to concern yourself with setting the offset of the record you're reading; this is handled by the consumer application itself. The following snippet writes out the offset, key, and message to the console:

```
for (ConsumerRecord<String, String> record : records)
        System.out.printf("offset = %d, key = %s, value = %s\n",
            record.offset(), record.key(), record.value());
```

The Final Code

When all this is put together, you have the basics of a stand-alone consumer application. The main components are setting the properties, creating the consumer, polling the broker for a batch of records to process, and then iterating the collection and working with each record.

```
import java.util.Properties;
import java.util.Arrays;
import org.apache.kafka.clients.consumer.KafkaConsumer;
import org.apache.kafka.clients.consumer.ConsumerRecords;
import org.apache.kafka.clients.consumer.ConsumerRecord;

public class ExampleConsumer {
   public static void main(String[] args) throws Exception {
        String topicName = "testtopic";
        Properties props = new Properties();

        props.put("bootstrap.servers", "localhost:9092");
        props.put("group.id", "testcg");
        props.put("enable.auto.commit", "true");
        props.put("auto.commit.interval.ms", "1000");
        props.put("session.timeout.ms", "30000");
        props.put("key.deserializer",
           "org.apache.kafka.common.serializa-tion.StringDeserializer");
        props.put("value.deserializer",
           "org.apache.kafka.common.serializa-tion.StringDeserializer");
        KafkaConsumer<String, String> consumer = new KafkaConsumer
           <String, String>(props);
```

```
consumer.subscribe(Arrays.asList(topicName))


System.out.println("Subscribed to topic " + topicName);
int i = 0;

while (true) {
    ConsumerRecords<String, String> records = consumer.poll(100);
    for (ConsumerRecord<String, String> record : records)


    System.out.printf("offset = %d, key = %s, value = %s\n",
        record.offset(), record.key(), record.value());
    }
  }
}
```

Building and Running the Applications

Now that you've seen how the applications are built, let's build them and run them. From the command line, run Maven to build the package.

```
$ mvn package
```

If that's successful, you'll see the path to the full JAR file with all the dependencies included.

```
[INFO] Building jar: /path/to/your/files/target/MLChapter12Kafka/target/
ch12kafka-jar-with-dependencies.jar
[INFO] ------------------------------------------------------------
[INFO] BUILD SUCCESS
[INFO] ------------------------------------------------------------
[INFO] Total time:  9.477 s
```

Start up your Kafka stand-alone cluster (if you need help, then you can find the full rundown earlier in this chapter). Once Zookeeper and Kafka are running, then you can execute the JAR file and produce and then consume some messages.

With everything running, open a new terminal window. Now it's time to run the applications in turn.

The Consumer Application

Before we send any messages to the cluster, let's get the consumer application running first. Open a terminal window and type in the following command from the project directory:

```
$ java -cp target/ch12kafka-jar-with-dependencies.jar mlbook.ch12.
examples.ExampleProducer
```

You'll see the application start and wait for messages.

```
$ java -cp target/ch12kafka-jar-with-dependencies.jar mlbook.ch12.
examples.ExampleConsumer
SLF4J: Failed to load class "org.slf4j.impl.StaticLoggerBinder".
SLF4J: Defaulting to no-operation (NOP) logger implementation
SLF4J: See http://www.slf4j.org/codes.html#StaticLoggerBinder for further
details.
Subscribed to topic testtopic
```

Now it's time to send some messages to the topic with the producer application.

The Producer Application

Open another terminal window separate from the consumer application. This application will send 10 messages to the testtopic topic.

```
$ java -cp target/ MLChapter12Kafka/target/ch12kafka-jar-with-
dependencies.jar mlbook.ch12.examples.ExampleProducer
```

The output will display that the messages have been sent.

```
$ java -cp target/ch12kafka-jar-with-dependencies.jar mlbook.ch12.
examples.ExampleProducer
SLF4J: Failed to load class "org.slf4j.impl.StaticLoggerBinder".
SLF4J: Defaulting to no-operation (NOP) logger implementation
SLF4J: See http://www.slf4j.org/codes.html#StaticLoggerBinder for further
details.
Messages sent successfully
```

Switch back to the consumer application and take a look at the output; it should look something like the following:

```
Subscribed to topic testtopic
offset = 0, key = 0, value = 0
offset = 1, key = 1, value = 1
offset = 2, key = 2, value = 2
offset = 3, key = 3, value = 3
offset = 4, key = 4, value = 4
offset = 5, key = 5, value = 5
offset = 6, key = 6, value = 6
offset = 7, key = 7, value = 7
offset = 8, key = 8, value = 8
offset = 9, key = 9, value = 9
```

The offset is the position in the Kafka topic log, the key is the given key of the message (that could be a UUID, for example), and the value is the actual message value. If I send more messages from the producer application (a simple case of rerunning the application), another 10 messages will be sent, and the consumer will poll the topic again. The output will show the new messages. If you take a look at the offset, you'll see that it has increased, but the key and values will be the same as the previous payload.

```
Subscribed to topic testtopic
offset = 0, key = 0, value = 0
offset = 1, key = 1, value = 1
offset = 2, key = 2, value = 2
offset = 3, key = 3, value = 3
offset = 4, key = 4, value = 4
offset = 5, key = 5, value = 5
offset = 6, key = 6, value = 6
offset = 7, key = 7, value = 7
offset = 8, key = 8, value = 8
offset = 9, key = 9, value = 9
offset = 10, key = 0, value = 0
offset = 11, key = 1, value = 1
offset = 12, key = 2, value = 2
offset = 13, key = 3, value = 3
offset = 14, key = 4, value = 4
offset = 15, key = 5, value = 5
offset = 16, key = 6, value = 6
offset = 17, key = 7, value = 7
offset = 18, key = 8, value = 8
offset = 19, key = 9, value = 9
```

If you were to close the consumer application and send more messages from the producer application another two times, when you start the consumer application, it will pick up from the last offset (position 20) and process the new 20 messages sent from the consumer. To quit from the consumer application, press Ctrl+C, and you will return to the command line.

The Streaming API

The consumer application processes messages. On the surface it's quite basic, and that's where the beauty lies with Kafka. If you need more power and features, then the streaming API is going to be a better bet.

With a streaming application, you have access to both the producer and consumer APIs and access to lambda-like functions that enable you to map, filter, and reduce on messages passing through the application. If you require aggregation functions like sliding and fixed windows and tabling, then the streaming API is the perfect platform to do it on.

The following code listing outlines the basics of the classic word count application. You'll see the properties configured the same as the producer and consumer applications already covered in this chapter. Within the listing, the core of the application is in the `createWordCountStream` method.

The KTable then converts each line into an array of lowercase words using a regular expression. The words are mapped and grouped by each word. At this point, there's no processing on the key as it's not required; it's just the word that we're after. Lastly, the groups are then counted and sent to the output topic.

```
static void createWordCountStream(final StreamsBuilder builder) {
       final KStream<String, String> lines = builder.stream(inputTopic);
       final Pattern pattern = Pattern.compile("\\W+", Pattern.UNICODE_
CHARACTER_CLASS);

       final KTable<String, Long> counts = lines
               .flatMapValues(value -> Arrays.asList(pattern.
split(value.toLowerCase())))
               .groupBy((keyIgnored, word) -> word)
               .count();
       counts.toStream().to(outputTopic, Produced.with(Serdes.String(),
Serdes.Long()));
    }
```

Streaming Word Counts

This application is moving forward from the old word count examples in batch processing from the time when the world was adopting Hadoop. The challenge back then was how to do that kind of processing in a real-time (or near-real-time) setting. The Kafka Streaming API gives you the tools to do this.

```
import org.apache.kafka.common.serialization.Serdes;
import org.apache.kafka.streams.KafkaStreams;
import org.apache.kafka.streams.StreamsBuilder;
import org.apache.kafka.streams.StreamsConfig;
import org.apache.kafka.streams.kstream.KStream;
import org.apache.kafka.streams.kstream.KTable;
import org.apache.kafka.streams.kstream.Produced;

import java.util.Arrays;
import java.util.Properties;
import java.util.regex.Pattern;

public class ExampleStreamingAPI {
    static final String inputTopic = "wcinput";
    static final String outputTopic = "wcoutput";

    static Properties getStreamsConfiguration(final String
bootstrapServers) {
       final Properties sConfig = new Properties();
       sConfig.put(StreamsConfig.APPLICATION_ID_CONFIG, "streaming-api-
example-app");
       sConfig.put(StreamsConfig.CLIENT_ID_CONFIG, "streaming-api-example-
client");
```

```
                sConfig.put(StreamsConfig.BOOTSTRAP_SERVERS_CONFIG,
        bootstrapServers);
                sConfig.put(StreamsConfig.DEFAULT_KEY_SERDE_CLASS_CONFIG, Serdes.
        String().getClass().getName());
                sConfig.put(StreamsConfig.DEFAULT_VALUE_SERDE_CLASS_CONFIG,
        Serdes.String().getClass().getName());
                return sConfig;
            }


            static void createWordCountStream(final StreamsBuilder builder) {
                final KStream<String, String> lines = builder.stream(inputTopic);
                final Pattern pattern = Pattern.compile("\\W+", Pattern.UNICODE_
        CHARACTER_CLASS);

                final KTable<String, Long> counts = lines
                        .flatMapValues(value -> Arrays.asList(pattern.
        split(value.toLowerCase())))
                        .groupBy((keyIgnored, word) -> word)
                        .count();
                counts.toStream().to(outputTopic, Produced.with(Serdes.String(),
        Serdes.Long())));
            }

            public static void main(final String[] args) {
                final String bootstrapServers = "localhost:9092";
                final Properties streamsConfiguration = getStreamsConfiguration(b
        ootstrapServers);
                final StreamsBuilder builder = new StreamsBuilder();
                createWordCountStream(builder);
                final KafkaStreams streams = new KafkaStreams(builder.build(),
        streamsConfiguration);
                streams.cleanUp();
                streams.start();
                Runtime.getRuntime().addShutdownHook(new Thread(streams::close));
            }
        }
```

Building a Streaming Machine Learning System

So far in this chapter you have learned about Kafka, the message log, the required code, and how it is all put together. Now it's time to turn this knowledge into a proof-of-concept machine learning system.

The remainder of this chapter concentrates on the machine learning aspects, including planning the system and putting the parts together. At the end of this chapter, you'll have a self-learning machine learning system built on Kafka.

To illustrate what's possible, I'll use a simple dataset to enable predictions. By creating different machine learning and statistical models (a decision tree, a multilayer perceptron, and linear regression), you have a system that can give predictions across all three models, over time one model may give a more robust prediction than the others.. On the surface, the data will look basic, but it will give you a framework for what's required if you want to combine streaming data with machine learning prediction.

Planning the System

To start, it's worth writing down what's required to acquire the data, train the models, and deal with predictions. Also consider such things as where you will persist the data and the models. Will they be on the file system? Or are you going to persist them on a storage service like Amazon S3 or on Azure's storage facility?

So, here's a quick list of questions to consider during the planning stage:

- What topics do we require?
- Where is the data coming from?
- What format is the data in?
- What algorithms are we going to use?
- How will the training be triggered?
- Where will the models be saved?
- How do we determine which models to use for predictions?
- How do we perform a prediction?

Figure 12.7 shows a sample plan.

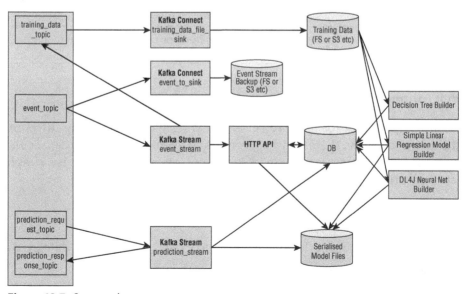

Figure 12.7: System plan

What Topics Do We Require?

This system is performing a couple of functions. First, there's data going into the system, so there needs to be a topic that's dedicated for events. This will be the first port of call for our event messages; we'll look at the actual message content shortly.

Event data comes in two forms, training data and commands, and will be handled by a streaming API application. If the event contains a command, then it should be executed within the application; otherwise, the event is training data, and the system will push that to another topic, the training data topic.

The training data topic is accepting the actual content, which is a line of comma-separated variables. There's no custom-made consumer application handling this information; this will be completed by using Kafka Connect, which will consume the CSV data for us and persist it to a file.

Predictions require two topics, one for the request and one for the response. The prediction request will take on a JSON payload with a key to request a model type to use and a feature value to make the prediction with.

The final breakdown looks like Table 12.3.

Table 12.3: Topic Breakdown

TOPIC NAME	DESCRIPTION
event _ stream	Incoming event messages with either a command or a training type.
training _ data _ topic	The topic handles raw CSV entries.
prediction _ request _ topic	When the user wants to run a prediction, it's via this topic. The JSON message will have a model type and a feature value to predict against.
prediction _ response _ topic	Results from the prediction are sent to this topic.

What Format Is the Data In?

The message format for this project is JSON. As we are handling different actions in each event, there may be a command event or a training data event. The prediction requests are also in JSON format. The reason for this is a simple matter of future proofing.

Events

Events can take one of two forms, a command event and a training data event.

```
{"type":"command", "payload":"build_mlp"}
```

The command event is basically an instruction to trigger a build of the machine learning model. They're not built all at the same time as that would start to drag down the system and degrade performance. So, the idea is that we can trigger any build type at any time. So, for example, `build _ mlp` will trigger the build for our neural network multilayer perceptron.

```
{"type":"training", "payload":"3,4,5,6"}
```

The training data just consists of the CSV values in the payload.

Predictions

To make a prediction via Kafka, the process is to send a JSON message to the prediction request topic. The payload contains the values of the three values from the scores, and the model returns the predicted value.

```
{"model":"mlp", "payload":"3,4,5"}
```

The responses from the model aren't returned in a traditional programming sense but sent as a message to the prediction response topic. It's up to the developer to create a consumer application to read the response from Kafka.

Continuous Training

There are numerous options for how to update our trained models. As data passes through the training topic and is persisted to disk through the connector, there needs to be some form of mechanism to trigger the training commands.

One approach is to send an event to the event topic itself. Once the Kafka streaming job parses the message, it will send a request to the API to start the required model build.

The alternative to sending a payload message is to use a scheduled cron job on the server. This does have one downside; it means that there's a command action that isn't registered in the event stream.

How to Install the Crontab Entries

If you want to install the cron entries, then you must do this via the `crontab` command. Open a terminal window and type the following:

```
crontab -e
```

Depending on the editor you have set as the default (it could be vi or nano, for example), the current crontab will be displayed.

Either type or copy and paste the following lines into the editor and then save the file:

```
# Crontab entry for Kafka DL4J MLP training.
#
# Daily run at 0100
# Run the decision tree builder
0 1 * * * /opt/mlbook/projects/dl4j.mlp/scripts/rundtr.sh > /var/log/
kafka-dl4j-training.log 2>&1
# Run the linear regression builder
30 1 * * * /opt/mlbook/projects/dl4j.mlp/scripts/runslr.sh > /var/log/
kafka-dl4j-training.log 2>&1
# Run the multi layer perceptron builder.
45 1 * * * /opt/mlbook/projects/dl4j.mlp/scripts/runmlp.sh > /var/log/
kafka-dl4j-training.log 2>&1
```

Once saved, `crontab` will schedule the model builds for the times shown in Table 12.4.

Table 12.4: Crontab Model Builds

MODEL NAME	BUILD TIME
Decision tree	01:00am
Simple linear regression	01:30am
Neural network	01:45am

Determining Which Models to Use for Predictions

We need to persist these results when they are performed. I'm going to use a relational database to save these results, and for this walk-through, I will use MySQL as it's commonly used and widely installed. If you are more comfortable using another database, then the schemas used can be easily modified.

As you have seen, when the models perform their training, each of the algorithms will output the results of the training. With the decision tree and the neural network, the model accuracy, the UUID of the model, and the training time are persisted in a table called `training _ log`.

The linear regression model is handled differently, as we are not saving a trained model to the filesystem, the outputs need to be persisted to the database. The slope and the intercept are persisted along with the UUID and training time in the `linear _ model` table.

When the Kafka streaming API jobs are running, they will refer to the training information in the tables to select the best models to run.

Setting Up the Database

The schema of the database is in the repository for this chapter. To create the database in MySQL, do each of the instructions from the command line.

```
$ mysqladmin -u root -p<your admin password> create mlchapter12
```

Replace `<your admin password>` with your admin password when you set up MySQL. If you don't have one, then you can remove the `-p` flag. Next run the MySQL client again and read in the `schema.sql` file.

The next step is to create the schema file. Open your text editor of choice and reproduce the schema shown here:

```
DROP TABLE IF EXISTS `linear_model`;
/*!40101 SET @saved_cs_client     = @@character_set_client */;
/*!40101 SET character_set_client = utf8 */;
CREATE TABLE `linear_model` (
  `id` int(11) NOT NULL AUTO_INCREMENT,
  `uuid` varchar(255) NOT NULL DEFAULT '',
  `createdate` datetime DEFAULT NULL,
  `slope` double DEFAULT NULL,
  `intercept` double DEFAULT NULL,
  `rsq` double DEFAULT NULL,
  `logoutput` text,
  PRIMARY KEY (`id`)
) ENGINE=InnoDB AUTO_INCREMENT=6 DEFAULT CHARSET=latin1;
/*!40101 SET character_set_client = @saved_cs_client */;


DROP TABLE IF EXISTS `training_log`;
/*!40101 SET @saved_cs_client     = @@character_set_client */;
/*!40101 SET character_set_client = utf8 */;
CREATE TABLE `training_log` (
  `id` int(11) NOT NULL AUTO_INCREMENT,
  `uuid` varchar(100) NOT NULL DEFAULT '',
  `training_date` datetime DEFAULT NULL,
  `train_eval_split` double NOT NULL DEFAULT '0.65',
  `execution_time` int(11) NOT NULL DEFAULT '-1',
  `model_accuracy` double NOT NULL DEFAULT '0',
  `training_output` text,
  `model_type` varchar(50) NOT NULL DEFAULT '',
  PRIMARY KEY (`id`)
) ENGINE=InnoDB AUTO_INCREMENT=14 DEFAULT CHARSET=latin1;
```

With that saved, you can now install it into MySQL. Run the following command from the same directory that you saved the schema file in:

```
$ mysql -u root -p<your admin password> mlchapter12 < schema.sql
```

With the database set up, you can now turn your attention to the models that are going to be used in the project.

Determining Which Algorithms to Use

So far in this book we've looked at a single set of data against a single type of algorithm in a "Well, it's this, so let's do this" manner. Depending on the data volumes, some algorithms may be better suited than others. With the streaming nature of the data, some algorithms may be more favorable than others.

For example, a J48 decision tree algorithm may be far more effective than a neural network to start off with, but over time it may be less effective for training with large volumes of data. What if we want to test different types of neural networks with different numbers of hidden nodes?

With this proof-of-concept project, we have the chance to try any model we want. For this chapter, I'm going to concentrate on three: the J48 decision tree built from Weka, which is an old but very effective Java-based machine learning library; a multilayer perceptron using DL4J; and, finally, a model using simple linear regression from the Apache Commons Math library.

The beauty of this system is that you can hold different model types in memory and predict against one or all of them. If you decide to add another model, like a support vector machine, then you can do so once you have the basic model coded.

Let's take a closer look at the three models I'm going to use.

Decision Trees

The Weka machine learning library has been around a few years now. For some personal projects, it's still my go-to library as the footprint is small. And for most things like CSV files, it's perfect. For a more comprehensive look at Weka, please refer to Chapter 5, "Working with Decision Trees."

The process for this model is simple. Let's walk through the steps.

Creating the J48 Instance

The main class for Weka-based decision trees is the J48 algorithm. Once the class has created the instances, the data is loaded in via a buffered reader stream.

With the data loaded, the next job is to is define the index of the data, which is the main output class. When predictions are made, it's this class that we're predicting. The last step is to then build the J48 classifier.

```
J48 cls = new J48();

Instances inst = new Instances(new BufferedReader(new FileReader("/opt/
mlbook/testdata/alldata.arff")));

inst.setClassIndex(inst.numAttributes() - 1);
cls.buildClassifier(inst);
```

Running the Evaluation

With the classifier built, we want to see how the model is performing. Running the evaluation will produce some resulting data that we'll persist in the training database.

```
Evaluation evaluation = new Evaluation(inst);
Random rand = new Random(1);
int folds = 10;
evaluation.crossValidateModel(cls, inst, folds, rand);
```

The cross-validation method takes the model, the instances, the number of folds (the number of sets to create from the training data to train and test against), and a random number seed.

Persisting the Model

The model can be serialized and persisted to the file system for reuse. I'm using a generated UUID that will also be written to the database so the streaming API job can retrieve the model later.

```
public void persistModel(J48 cls, String uuid) {
    try {
      SerializationHelper.write("/path/to/models/" + uuid+ ".model", cls);
    } catch (Exception e) {
      e.printStackTrace();
    }
}
```

Updating the Database

With the model persisted, the evaluation data and the time taken to create the model can be written to the database. The main metric I'm interested in is the accuracy of the model, and the `pctCorrect()` method in the evaluation class will produce this information. There is a class within the project called `DBTools.java` that acts as the helper class to update and retrieve the model metadata from the models when they are built and also when predictions are made.

```
DBTools.writeResultsToDBuuid, -1, (stop - start), evaluation.
pctCorrect()/100, evaluation.toSummaryString(), "dtr");
```

The Final Code

This is the full code listing for the J48 decision tree model generator. Essentially this is a stand-alone application, as are the other two models. So, this can be run in isolation as a generator for decision tree applications.

```java
package mlbook.ch12.kafka.mlp;

import weka.classifiers.Evaluation;
import weka.classifiers.trees.J48;
import weka.core.Instances;
import weka.core.SerializationHelper;
import weka.gui.treevisualizer.PlaceNode2;
import weka.gui.treevisualizer.TreeVisualizer;

import java.awt.*;
import java.io.BufferedReader;
import java.io.FileNotFoundException;
import java.io.FileReader;
import java.util.Random;
import java.util.UUID;

public class DecisionTreeBuilder {

    public DecisionTreeBuilder() {

        String uuid = UUID.randomUUID().toString();


        // build the classifier
        J48 cls= buildModel(uuid);

        // persist the model to the file system
        persistModel(cls, uuid);


    }

    public J48 buildModel(String uuid) {

        //Classifier cls = new J48();
        J48 cls = new J48();
        try {
            Instances inst = new Instances(new BufferedReader(new
FileReader("/opt/mlbook/testdata/alldata.arff")));
            inst.setClassIndex(inst.numAttributes() - 1);
            try {
                long start = System.currentTimeMillis();
                cls.buildClassifier(inst);

                Evaluation evaluation = new Evaluation(inst);
                Random rand = new Random(1);
                int folds = 10;

                evaluation.crossValidateModel(cls, inst, folds, rand);
                long stop = System.currentTimeMillis();

                DBTools.writeResultsToDB(uuid, -1, (stop - start),
evaluation.pctCorrect()/100, evaluation.toSummaryString(), "dtr");
```

```
        } catch (Exception e) {
            e.printStackTrace();
        }
    } catch (FileNotFoundException e) {
        e.printStackTrace();
    } catch (Exception e) {
        e.printStackTrace();
    }

    return cls;

}

public void persistModel(J48 cls, String uuid) {
    try {
        SerializationHelper.write("/opt/mlbook/testdata/models/" + uuid+
".model", cls);
    } catch (Exception e) {
        e.printStackTrace();
    }
}

public static void main(String[] args) {
    DecisionTreeBuilder ddt = new DecisionTreeBuilder();
}
}
```

Simple Linear Regression

There are times when linear regression beats everything else. As this proof of concept is with basic data, it makes sense to put this model into the mix too. Part of this code is covered in Chapter 4.

Creating the Model

Using the Apache Commons Math library, we build a linear regression with the SimpleRegression class. As each line of the data file are strings, each line needs splitting (on the comma), resulting in a string array (a String[] type). I'm using the parseDouble() function of the Double class to parse the first and second elements of the string array.

With the two double values, they are added to the regression model. With every line added, the model is returned.

```
SimpleRegression sr = new SimpleRegression();
    for(String s : lines) {
      String[] ssplit = s.split(",");
      double x = Double.parseDouble(ssplit[0]);
      double y = Double.parseDouble(ssplit[1]);
      sr.addData(x,y);
    }
return sr;
```

Evaluating

There's no actual model evaluation as with the neural network or the decision tree. The SimpleRegression class does give some output in terms of the slope, intercept values, the standard error value, and the R2 value to see the variance of the data covered in the regression. The adjusted R2 value is used as the rank of the model in the database.

As a final step to prove the model is working, the program will run some predictions against the model. It shows the input score (a random number) and the prediction from the regression model.

All this output is returned as a string.

```
sb.append("Intercept: " + sr.getIntercept());
        sb.append("\n");
        sb.append("SloComp: " + sr.getSlope());
        sb.append("Standard Error: " + sr.getSlopeStdErr());
        sb.append("Adjusted R2 value: " + sr.getRSquare());

Random r = new Random();
        for (int i = 0 ; i < runs ; i++) {
            int rn = r.nextInt(10);
            sb.append("Input score: " + rn + " prediction: " + Math.
round(sr.predict(rn)));
            sb.append("\n");
        }
```

Saving the Model to the Database

There's no model to save as such like we do with the neural network and the decision tree. All we have to do is persist the values into the database. The slope and intercept are required to make future predictions against. We also write the R2 value for our accuracy ranking. The time it took to create the model and also the evaluation that we created are also persisted.

```
DBTools.writeLinearResults(uuid, sr.getIntercept(), sr.getSlope(),
sr.getRSquare(), time, runPredictions(sr, 20));
```

The Final Code

The code loads the data, creates the model, and persists the results to the database. There's no model to serialize. It's just the slope and intercept being saved for use later when the streaming API requests it.

```
package mlbook.ch12.kafka.mlp;

import org.apache.commons.math3.stat.regression.SimpleRegression;

import java.io.*;
import java.util.ArrayList;
```

```java
import java.util.List;
import java.util.Random;
import java.util.UUID;

public class LinearRegressionBuilder {

    private static String path = "/opt/mlbook/testdata/alloutput.csv";

    public LinearRegressionBuilder() {
        List<String> lines = loadData(path);
        long start = System.currentTimeMillis();
        SimpleRegression sr = getLinearRegressionModel(lines);

        long stop = System.currentTimeMillis();
        long time = stop - start;

        String uuid = UUID.randomUUID().toString();
        DBTools.writeLinearResults(uuid, sr.getIntercept(),
sr.getSlope(), sr.getRSquare(), time, runPredictions(sr, 20));
        runPredictions(sr, 40);
    }

    private SimpleRegression getLinearRegressionModel(List<String>
lines) {
        SimpleRegression sr = new SimpleRegression();
        for(String s : lines) {
            String[] ssplit = s.split(",");
            double x = Double.parseDouble(ssplit[0]);
            double y = Double.parseDouble(ssplit[1]);
            sr.addData(x,y);
        }

        return sr;
    }

    private String runPredictions(SimpleRegression sr, int runs) {
        StringBuilder sb = new StringBuilder();
        // Display the intercept of the regression
        sb.append("Intercept: " + sr.getIntercept());
        sb.append("\n");
        // Display the slope of the regression.
        sb.append("SloComp: " + sr.getSlope());
        sb.append("\n");
        // Display the slope standard error
        sb.append("Standard Error: " + sr.getSlopeStdErr());
        sb.append("\n");
        // Display adjusted R2 value
        sb.append("Adjusted R2 value: " + sr.getRSquare());
        sb.append("\n");
        sb.append("****************************************************");
        sb.append("\n");
```

```java
            sb.append("Running random predictions......");
            sb.append("\n");
            sb.append("");
            Random r = new Random();
            for (int i = 0 ; i < runs ; i++) {
                int rn = r.nextInt(10);
                sb.append("Input score: " + rn + " prediction: " + Math.
round(sr.predict(rn)));
                sb.append("\n");
            }
            return sb.toString();
        }

    private List<String> loadData (String filename) {
        List<String> lines = new ArrayList<String>();
        try {
            FileReader f = new FileReader(filename);
            BufferedReader br;
            br = new BufferedReader(f);
            String line = "";
            while ((line = br.readLine()) != null) {
                lines.add(line);
            }
        } catch (FileNotFoundException e) {
            System.out.println("File not found.");
        } catch (IOException e) {
            System.out.println("Error reading file");
        }

        return lines;
    }

    public static void main(String[] args) {
        LinearRegressionBuilder dlr = new LinearRegressionBuilder();
    }

}
```

Neural Network

If I'm totally honest, running a neural network against the current dataset probably isn't the best idea. The quantity of data isn't going to be enough to get any accurate results. . .yet. Planning for the future, though, is a good idea, and you want a model that can work with the unknown, so having this neural network in place is worthwhile.

Working on the assumption that data will be continuously streaming in via Kafka, then there's a strong case for having a neural network over the long term. So, I'm going to build a multilayer perceptron to handle the scoring data.

With things being so variable over time, it does raise some interesting questions on how the model is going to be built. With simple linear regression and decision trees the training and model require no configuration or planning: the training is run against the data, the model is created, and the evaluation is performed. A neural network, on the other hand, needs a little more crafting to get the best performance out of it.

Data Importing

I'm using the DeepLearning4J library to create this neural network application. One of the positive aspects of DL4J is the amount of thought that's gone into loading the data.

To import the CSV data, there's a dedicated `CSVRecordReader` class that will import the data based on a passed-in delimiter. As there is no header record, there are no lines to skip, so that value is set to zero.

```
int numLinesToSkip = 0;
String delimiter = ",";
RecordReader recordReader = new CSVRecordReader(numLinesToSkip,d
elimiter);

recordReader.initialize(new FileSplit(new File("/opt/mlbook/
testdata/")));
```

Hidden Nodes

How many hidden nodes should we use in our model? Given that the volume of data is going to increase as time goes on, it's going to be difficult to know how many hidden nodes to have.

There are a few scenarios for training our neural network. We could train against the full dataset as set intervals (more on that shortly) or, if the data volumes are huge, split out random datasets and train on those, creating several potential models to use.

With so many permutations of training data, it's hard to say how many hidden nodes the neural network would require. There is, however, a method to give a rough calculation of how many hidden nodes to use (see Figure 12.8).

$$\frac{samples}{S * (i + o)}$$

Figure 12.8: Calculation for hidden nodes

```
private static long getHiddenNodeEstimate(int inputs, int outputs, int
samplesize) {
        Random r = new Random();
        double out  = (samplesize / ((inputs + outputs) * r.nextInt(9) +
1));
        return Math.round(out);
    }
```

The previous code takes the number of input nodes, the number of output nodes, and the sample size (number of lines in a CSV file, for example). For the calculation to take place, we need to add some randomness so there's a random number introduced into the equation. The result of the method is the number of hidden nodes to use. Over time, with more training, you will build up a picture of the sweet spot of hidden nodes. It's not an exact science, but it's something that needs to be repeated so we get the most accurate model.

When we talk about neural networks, it tends to be about a fixed set of parameters. There are x inputs, y outputs, and z hidden nodes. The data tends to be a fixed quantity with no forward planning and rerunning. Data, however, is ever evolving.

Model Configuration

The configuration for the neural network is a simple multilayer perceptron. I'm using four layers: an input layer, output layer, and two hidden layers. Deciding on how many hidden nodes are in the hidden layers comes down to the result of the getHiddenNodesEstimate method, as discussed earlier.

```
MultiLayerConfiguration conf = new NeuralNetConfiguration.Builder()
                .seed(seed)
                .iterations(iterations)
                .activation(Activation.TANH)
                .weightInit(WeightInit.XAVIER)
                .learningRate(0.1)
                .regularization(true).l2(1e-4)
                .list()
                .layer(0, new DenseLayer.Builder().nIn(numInputs).
nOut(hiddenNodes).build())
                .layer(1, new DenseLayer.Builder().nIn(hiddenNodes).
nOut(hiddenNodes).build())
                .layer(2, new DenseLayer.Builder().nIn(hiddenNodes).
nOut(hiddenNodes).build())
                    .layer(3, new OutputLayer.Builder(LossFunctions.
LossFunction.NEGATIVELOGLIKELIHOOD)
                    .activation(Activation.SOFTMAX)
                    .nIn(hiddenNodes).nOut(outputNum).build())
                .backprop(true).pretrain(false)
                .build();
```

Model Training

With the configuration completed, the model is then created. The training split is set for 65 percent of the training data, with the remaining 35 percent being used for evaluation. As the model is being trained, the application will output the progress every 100 iterations; the training of the model is set in the application at 2,000 iterations.

```
MultiLayerNetwork model = new MultiLayerNetwork(conf);
model.init();
model.setListeners(new ScoreIterationListener(100));

model.fit(trainingData);
```

Evaluation

The evaluation steps don't require any intervention from the application itself; it's more a reporting exercise. With our 11 output nodes, the evaluation class generates an output based on the feature matrix of the test data. Once the evaluation is complete, it's output to the console.

```
Evaluation eval = new Evaluation(11);
        log.info("Getting evaluation");
        INDArray output = model.output(testData.getFeatureMatrix());
        log.info("Getting evaluation output");
        eval.eval(testData.getLabels(), output);
        System.out.println(eval.stats());
```

Saving the Model Results to the Database

As with the other models, the evaluation results are stored in the database. A figure of the split of training data is taken, along with the time taken to create the model, and the final F1 score of the model. When the prediction events happen, the criteria on what model will be used will be determined by the F1 score.

```
// Write output results to database
        DBTools.writeResultsToDB(uuid, evalsplit, timetaken, eval.f1() ,
    eval.stats(), "mlp");
```

Persisting the Model

Models generated in DeepLearning4Java are persisted as ZIP files. The ModelSerializer class gives us an easy way to persist the generated model. In our demo system, there's no need to update models; we are only creating new models each run, so we set the saveUpdater Boolean value to false.

In this example, we're passing the generated UUID as the filename of the ZIP file. This UUID will be the same one persisted in the database table.

```
            // Save model
            File locationToSave = new File("/opt/mlbook/testdata/models/" +
uuid + ".zip");
            boolean saveUpdater = false;
            ModelSerializer.writeModel(model, locationToSave, saveUpdater);
```

The Final Code

Here's the full code for the neural network. This includes all the steps that I've walked through and also implements CSVRecordReader, which is a helper class to parse the CSV file.

This model also implements the hidden node method, taking the input and output nodes plus the number of samples in the dataset.

```
package mlbook.ch12.kafka.mlp;

import org.datavec.api.records.reader.RecordReader;
import org.datavec.api.records.reader.impl.csv.CSVRecordReader;
import org.datavec.api.split.FileSplit;
import org.deeplearning4j.datasets.datavec.RecordReaderDataSetIterator;
import org.deeplearning4j.eval.Evaluation;
import org.deeplearning4j.nn.conf.MultiLayerConfiguration;
import org.deeplearning4j.nn.conf.NeuralNetConfiguration;
import org.deeplearning4j.nn.conf.layers.DenseLayer;
import org.deeplearning4j.nn.conf.layers.OutputLayer;
import org.deeplearning4j.nn.multilayer.MultiLayerNetwork;
import org.deeplearning4j.nn.weights.WeightInit;
import org.deeplearning4j.optimize.listeners.ScoreIterationListener;
import org.deeplearning4j.util.ModelSerializer;
import org.nd4j.linalg.activations.Activation;
import org.nd4j.linalg.api.ndarray.INDArray;
import org.nd4j.linalg.dataset.DataSet;
import org.nd4j.linalg.dataset.SplitTestAndTrain;
import org.nd4j.linalg.dataset.api.iterator.DataSetIterator;
import org.nd4j.linalg.dataset.api.preprocessor.DataNormalization;
import org.nd4j.linalg.dataset.api.preprocessor.NormalizerStandardize;
import org.nd4j.linalg.lossfunctions.LossFunctions;
import org.slf4j.Logger;
import org.slf4j.LoggerFactory;

import java.io.File;
import java.sql.Connection;
import java.sql.DriverManager;
import java.sql.PreparedStatement;
import java.sql.SQLException;
import java.util.Random;
import java.util.UUID;
```

```
public class ANNBuilder {

    private static Logger log = LoggerFactory.getLogger(ANNBuilder.class);


    private static long getHiddenNodeEstimate(int inputs, int outputs,
int samplesize) {
        Random r = new Random();
        double out  = (samplesize / ((inputs + outputs) * r.nextInt(9) +
1));
        return Math.round(out);
    }

    public static void main(String[] args) throws  Exception {
    // Everything is classed as a new run so we want a UUID for each
model run.

        String uuid = UUID.randomUUID().toString();
    long start = System.currentTimeMillis();
        //First: get the dataset using the record reader.
CSVRecordReader handles loading/parsing
        int numLinesToSkip = 0;
        String delimiter = ",";
        RecordReader recordReader = new CSVRecordReader(numLinesToSkip,
delimiter);

        recordReader.initialize(new FileSplit(new File("/opt/mlbook/
testdata/")));


        //Second: the RecordReaderDataSetIterator handles conversion to
DataSet objects, ready for use in neural network
        int labelIndex = 3;     //4 values in each row of the CSV: 3
input features followed by an integer label (class) index.
        int numClasses = 11;    //11 classes
        int batchSize = 474;
        double evalsplit = 0.65;

        DataSetIterator iterator = new RecordReaderDataSetIterator(record
Reader,batchSize,labelIndex,numClasses);

        DataSet allData = iterator.next();
        allData.shuffle();
        SplitTestAndTrain testAndTrain = allData.
splitTestAndTrain(evalsplit);  //Use 65% of data for training

        DataSet trainingData = testAndTrain.getTrain();
        DataSet testData = testAndTrain.getTest();

        //We need to normalize our data. We'll use NormalizeStandardize
(which gives us mean 0, unit variance):
```

```
        DataNormalization normalizer = new NormalizerStandardize();
        normalizer.fit(trainingData);
        normalizer.transform(trainingData);
        normalizer.transform(testData);


        final int numInputs = 3;
        int outputNum = 11;
        int iterations = 2000;
        long seed = 6;

        int hiddenNodes = (int)getHiddenNodeEstimate(numInputs,
outputNum, batchSize);



        log.info("Build model....");
        MultiLayerConfiguration conf = new NeuralNetConfiguration.
Builder()
                .seed(seed)
                .iterations(iterations)
                .activation(Activation.TANH)
                .weightInit(WeightInit.XAVIER)
                .learningRate(0.1)
                .regularization(true).l2(1e-4)
                .list()
                .layer(0, new DenseLayer.Builder().nIn(numInputs).
nOut(hiddenNodes).build())
                .layer(1, new DenseLayer.Builder().nIn(hiddenNodes).
nOut(hiddenNodes).build())
                .layer(2, new DenseLayer.Builder().nIn(hiddenNodes).
nOut(hiddenNodes).build())
                    .layer(3, new OutputLayer.Builder(LossFunctions.
LossFunction.NEGATIVELOGLIKELIHOOD)
                    .activation(Activation.SOFTMAX)
                    .nIn(hiddenNodes).nOut(outputNum).build())
                .backprop(true).pretrain(false)
                .build();

        //run the model
        MultiLayerNetwork model = new MultiLayerNetwork(conf);
        model.init();
        model.setListeners(new ScoreIterationListener(100));

        model.fit(trainingData);

        log.info("Made it here.....");
        long stop = System.currentTimeMillis();
        long timetaken = stop - start;
```

```
        System.out.println("Took " + timetaken + " millis");

        //evaluate the model on the test set
        Evaluation eval = new Evaluation(11);
        log.info("Getting evaluation");
        INDArray output = model.output(testData.getFeatureMatrix());
        log.info("Getting evaluation output");
        eval.eval(testData.getLabels(), output);
        System.out.println(eval.stats());

        // Write output results to database
        DBTools.writeResultsToDB(uuid, evalsplit, timetaken, eval.f1() ,
eval.stats(), "mlp");

        // Save model
        File locationToSave = new File("/opt/mlbook/testdata/models/" +
uuid + ".zip");
        boolean saveUpdater = false;
        ModelSerializer.writeModel(model, locationToSave, saveUpdater);

    }

}
```

Kafka Topics

In this application, there are four topics. First, there's the event stream topic; this contains the messages for either the training data or the commands to run a build on one of the three models we have. If the message contains training data, the streaming application will push the contents of the message, the CSV data, to the training data topic.

With the commands and training data taken care of, it's just the predictions that we need to think about. To make a prediction, messages are sent to the prediction request topic. This is another JSON payload with the model type to run and the data required to make a prediction. The prediction response topic is a topic that the results are sent to. The streaming application handling the predictions publishes the result responses to this topic. For this proof of concept, I'm going to assume that there's a consumer handling this.

Creating the Topics

The topic creation is done using the `kafka-topics` command that was covered earlier in the chapter. I've created a shell script that creates all four topics for you. There are three pieces of information the topic command requires: the hostname and port for Zookeeper, the replication factor, and the number of partitions.

For learning purposes, I've kept the replication factor and partitions to 1. If this system were put into a production setting, then I would be looking at increasing those numbers to give some resilience of the overall application. The following is the full shell script to create the topics required:

```
KAFKA_HOME=/usr/local/kafka_2.11-1.0.0/
ZK_CONNECT=localhost:2181
REPLICATION_FACTOR=1
PARTITIONS=1


# training_data_topic - Where the raw data will be transported for
training. Using Kafka Connect to push the data to a filestore.
echo "Creating topic: training_data_topic"

$KAFKA_HOME/bin/kafka-topics.sh --create --topic training_data_topic
--zookeeper $ZK_CONNECT --partitions $PARTITIONS --replication-factor
$REPLICATION_FACTOR

# event_stream_topic - Events are sent to the Events Kafka Streaming API
application.
echo "Creating topic: event_topic"
$KAFKA_HOME/bin/kafka-topics.sh --create --topic event_topic --zookeeper
$ZK_CONNECT --partitions $PARTITIONS --replication-factor $REPLICATION_
FACTOR

# prediction_request_topic - Sends data to the Prediction Streaming
application.
echo "Creating topic: prediction_request_topic"
$KAFKA_HOME/bin/kafka-topics.sh --create --topic prediction_request_
topic --zookeeper $ZK_CONNECT --partitions $PARTITIONS --replication-
factor $REPLICATION_FACTOR

# prediction_response_topic - The prediction response from the original
prediction_request_topic
echo "Creating topic: prediction_response_topic"
$KAFKA_HOME/bin/kafka-topics.sh --create --topic prediction_response_
topic --zookeeper $ZK_CONNECT --partitions $PARTITIONS --replication-
factor $REPLICATION_FACTOR
```

Assuming Zookeeper and Kafka are running, open a terminal window and go to the directory with the script and execute it. If you're running Kafka as the root user, then you will need to run the script as root too.

```
sudo ./create_topics.sh
```

It's wise to confirm that the topics have been created, so run the Kafka topics command again to list the topics.

```
./kafka-topics --list --zookeeper localhost:2181
```

Kafka Connect

Within the Kafka ecosystem is Kafka Connect, which is basically a consumer application that can act either as a source, consuming data from an application, or as a sink, persisting or sending data to an application.

For this application there are two Connect sinks in operation—one to back up all the messages passing through the event topic and the other that persists the training data for the models to train against.

Why Persist the Event Data?

As you have seen throughout this chapter, the Kafka system is an immutable message log. Brokers have configurations, and as you'd expect, one of these settings is log retention, in other words, how long messages within the cluster are kept to be available to consumers.

Once the log retention threshold passes for a message, then it's marked for deletion and won't be available to any consumers depending on it. For most things, this is no issue. For this proof of concept, though, I may want to replay messages from the very start, including commands.

The reason for creating the event data sink, even though it's not used by an application, is to create a record of each data event in case we ever need to run it from scratch. If that needs to happen, then all that's required is to play the JSON payloads through the event _ topic again. It would be prudent to reset all the models and start from scratch. It gives a mechanism for disaster recovery.

Persisting Event Data

The event data passes through the event topic. While the actual payload is handled by the streaming application, it's prudent to back up the message data too. This Connect configuration persists the JSON payloads to a file called events. json. There's no transformation happening on the actual JSON messages; they are just appended to the file.

The Connect configuration requires a name and a topic to read from. The key and value converters will deserialize the data so it can be persisted safely. Connect requires a file path to write the data to.

With the configuration options put together, the final Connect configuration looks like this:

```
name=dl4j-eventsw-file-sink
connector.class=FileStreamSink
tasks.max=1
file=/opt/mlbook/testdata/events/events.json
topic=event_topic
key.converter=org.apache.kafka.connect.storage.StringConverter value.
converter=org.apache.kafka.connect.storage.StringConverter
```

Persisting Training Data

The training data sink works in the same way as the event topic backup sink. Changing the topic and the output file path are the only actions required. All that's being written to the file is CSV data.

```
name=dl4j-training-data-file-sink
connector.class=FileStreamSink
tasks.max=1
file=/opt/mlbook/testdata/connect/trainingdata.csv
topics=training_data_topic
key.converter=org.apache.kafka.connect.storage.StringConverter value.
converter=org.apache.kafka.connect.storage.StringConverter
```

Installing the Connector Configurations

The Kafka Connect scripts come in two types—one for stand-alone operation and the other for a distributed cluster. To start Kafka Connect on the stand-alone development cluster, you are required to pass in the Connect properties first and then the subsequent properties files of your connectors.

```
For this project example the following would be run (as one command) on
the command line.
 # bin/connect-standalone.sh config/connect-standalone.properties \
/path/to/repo/config/dl4j_event_to_fs_sink.properties \
/path/to/repo/config/dl4j_to_fs_sink.properties
```

Kafka will load in the required plugins and start Connect. If there is an error with the properties file, Connect will close, and you will return to the command line. When that happens, check your properties files for any errors and try again.

When Connect starts up correctly, you should see something like the following output. The main thing to look for are things like "Sink task finished initialization and start," meaning the connector has installed and is waiting.

```
[2019-08-11 14:36:38,032] INFO WorkerSinkTask{id=dl4j-eventsw-file-
sink-0} Sink task finished initialization and start (org.apache.kafka.
connect.runtime.WorkerSinkTask:301)
[2019-08-11 14:36:38,078] INFO Cluster ID: eh574OwJRguX33YZfyyVgg
(org.apache.kafka.clients.Metadata:365)
[2019-08-11 14:36:38,853] INFO [Consumer clientId=consumer-1,
groupId=connect-dl4j-eventsw-file-sink] Discovered group coordinator
192.168.1.102:9092 (id: 2147483647 rack: null) (org.apache.kafka.
clients.consumer.internals.AbstractCoordinator:675)
[2019-08-11 14:36:38,856] INFO [Consumer clientId=consumer-1,
groupId=connect-dl4j-eventsw-file-sink] Revoking previously assigned
partitions [] (org.apache.kafka.clients.consumer.internals.
ConsumerCoordinator:459)
[2019-08-11 14:36:38,856] INFO [Consumer clientId=consumer-1,
groupId=connect-dl4j-eventsw-file-sink] (Re-)joining group (org.apache.
kafka.clients.consumer.internals.AbstractCoordinator:491)
```

```
[2019-08-11 14:36:38,894] INFO [Consumer clientId=consumer-1,
groupId=connect-dl4j-eventsw-file-sink] (Re-)joining group (org.apache.
kafka.clients.consumer.internals.AbstractCoordinator:491)
[2019-08-11 14:36:38,988] INFO [Consumer clientId=consumer-1,
groupId=connect-dl4j-eventsw-file-sink] Successfully joined group
with generation 1 (org.apache.kafka.clients.consumer.internals.
AbstractCoordinator:455)
[2019-08-11 14:36:38,991] INFO [Consumer clientId=consumer-1,
groupId=connect-dl4j-eventsw-file-sink] Setting newly assigned
partitions: event_topic-0 (org.apache.kafka.clients.consumer.internals.
ConsumerCoordinator:290)
[2019-08-11 14:36:39,024] INFO [Consumer clientId=consumer-1,
groupId=connect-dl4j-eventsw-file-sink] Resetting offset for partition
event_topic-0 to offset 0. (org.apache.kafka.clients.consumer.internals.
Fetcher:584)
```

The REST API Microservice

The backbone of the system is a prediction HTTP API. This accepts requests from the two streaming applications and can also accept requests directly to the endpoint, which could be via a web page using an asynchronous Ajax call, for example.

Using the CompojureAPI library, it's a fairly trivial matter to create REST-based APIs that also include the Swagger front end, which makes testing a lot easier.

There are number of handlers that need to be implemented. Table 12.5 breaks down the endpoints, any required values, and the function of the endpoint.

Table 12.5: Breakdown of Endpoints

ENDPOINT	PARAMETERS	FUNCTION
/api/build _ dtr	None	Triggers the build script for decision tree model
/api/build _ slr	None	Triggers the build script for simple linear regression
/api/build _ mlp	None	Triggers the build script for the multilayer perceptron neural network build
/api/predict/dtr/:d	Three-part CSV line	Predicts the score against the decision tree model
/api/predict/slr/:d	Three-part CSV line	Predicts the score against the simple linear regression model
/api/predict/mlp/:d	Three-part CSV line	Predicts the score against the multilayer perceptron neural network

Using the framework to compose the API, there's little in the way of coding to do. Each function is essentially a GET handler that performs a function and returns a JSON payload as a response.

```
(GET "predict/dtr/:d" []
                  :path-params [d :- String]
                  :summary "Runs a prediction against the decision tree
model"
                  (ok (json/write-str (dtr/predict-decision-tree d))))
```

The full code for the REST API handler is shown here:

```
(ns prediction.http.api.handler
  (:require [compojure.api.sweet :refer :all]
            [ring.util.http-response :refer :all]
            [prediction.http.api.linear :as slr]
            [prediction.http.api.decisiontree :as dtr]
            [prediction.http.api.mlp :as mlp]
            [clojure.data.json :as json]
            [schema.core :as s]))

;; A very basic API to build and predict against the models.
;; Swagger interface is built as well so you can test.
(def app
  (api
    {:swagger
     {:ui "/"
      :spec "/swagger.json"
      :data {:info {:title "Prediction.http.api"
                    :description "Strata 2018 - Kafka/DL4J/Weka/Commons Demo"}
             :tags [{:name "api", :description "prediction api"}]}}}

    (context "/api" []
             :tags ["api"]

             (GET "/build_dtr" []
                  :return {:result String}
                  :summary "Builds the decision tree model"
                  (ok {:result (dtr/run-model-build)}))

             (GET "/build_mlp" []
                  :return {:result String}
                  :summary "Builds the neural network model"
                  (ok {:result (mlp/run-model-build)}))

             (GET "/build_slr" []
                  :return {:result String}
                  :summary "Builds the simple linear regression model"
                  (ok {:result (slr/run-model-build)}))
```

```
        (GET "/predict/mlp/:d" []
              :path-params [d :- String]
              :summary "Runs a prediction against the neural network
model"
              (ok (json/write-str (mlp/predict-mlp d))))

        (GET "predict/dtr/:d" []
              :path-params [d :- String]
              :summary "Runs a prediction against the decision tree
model"
              (ok (json/write-str (dtr/predict-decision-tree d))))

        (GET "/predict/slr/:d" []
              :path-params [d :- String]
              :summary "Runs a prediction against simple linear
regression models"
              (ok (json/write-str (slr/predict-simple-linear d)))))))
```

The Swagger interface gives you an accessible browser-based page where you can test the API and run commands and see the API responses (see Figure 12.9).

Figure 12.9: Swagger interface to test the API

Processing Commands and Events

The core of this system revolves around two Kafka streaming applications. One processes the commands and events coming in from the event topic, and the other streams the application to handle the predictions. Figure 12.10 shows the flow of how a message is mapped.

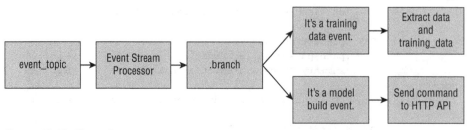

Figure 12.10: Flow of a message

I decided to write these in Clojure, mainly for ease of testing. When functions are small, they can be easily tested. When working with the REPL, it's easy to build out functions quickly in the REPL and then transfer them to your main codebase.

This streaming application uses the Java APIs to set up the streaming job, like the Java example you saw earlier in the chapter. There are a few features that I've added, though, to make starting the application and configuration easier.

Finding Kafka Brokers

As you've realized from using the Kafka command-line tools, there's a lot of specifying where broker lists and Zookeeper servers reside. After a while, it becomes hard to find and configure all of them, especially when it's a large cluster.

Within this application, there are two functions that will create the broker list for us at runtime, so there's no need for configuration or hard coding. The base Zookeeper server address is within the EDN configuration file.

```
(defn get-broker-list
  [zk-conf]
  (let [c (merge (zk-defaults/zk-client-defaults) zk-conf)]
    (with-open [u (client/make-zk-utils c false)]
      (cluster/all-brokers u))))

(defn broker-str [zkconf]
  (let [zk-brokers (get-broker-list zkconf)
        brokers (map (fn [broker] (str (get-in broker [:endpoints :
plaintext :host]) ":" (get-in broker [:endpoints :plaintext :port]))) )
zk-brokers)]
    (if (= 1 (count brokers))
      (first brokers)
      (cstr/join "," brokers)))))
```

The first function, `get-broker-list`, takes the Zookeeper client and returns all the brokers from the Zookeeper nodes. This is returned as a map; the `broker-str` function maps through the entries and retrieves the host name and port. The final step is to return the string. Depending on how many broker names exist, this is either a single entry or a comma-separated list of broker addresses.

A Command or an Event?

When a message passes through the event topic, the application needs to determine whether the payload is a training event containing a line of CSV data or a command.

Within the streaming API, there's a method to branch the stream. By testing on a condition, in this case whether our payload key is a command or a training line, the message values can be directed to the correct element of an array. This partitioned stream can then be processed correctly by accessing the position of the array.

For this example, the first array element consists of the commands; these are handled by the `process-command` function. At the processing of the command, the model type is taken and set to the API to start a model training job.

The training data messages are handled by the `process-training` function, and the CSV data is extracted from the payload and is returned. The stream then acts as a producer sending the raw data to the training data topic. Data going to this topic is persisted via the Kafka Connect job and appended to the training data file on the filesystem.

```
(do
  (let [partitioned-stream
        (.branch
         (.stream builder input-topic)
         (into-array Predicate [(reify
                                  Predicate
                                  (test [_ _ v]
                                    (do
                                      (log/info "p0 " v)
                                      (->> v
                                           (pre-process-data)
                                           (pred-key-type
"command")))))
                                (reify
                                  Predicate
                                  (test [_ _ v]
                                    (do
                                      (log/info "p1 " v)
                                      (->> v
                                           (pre-process-data)
                                           (pred-key-type
"training")))))])]
```

```
              training-topic-stream (.stream builder training-topic)]
        (log/info partitioned-stream)
        (log/info training-topic-stream)
        (-> (aget partitioned-stream 0)
            (.mapValues (reify ValueMapper (apply [_ v] (process-command
v))))
            (.print))
        (-> (aget partitioned-stream 1)
            (.mapValues (reify ValueMapper (apply [_ v] (process-
training v))))
            (.to training-topic-name)))
```

The final code listing takes the previous elements and also has the deserialization function to convert from a byte array to a string. Configuration is stored in an EDN file, which is basically a map of keys and values.

```
(ns kafka.stream.events.core
  (:require [franzy.admin.zookeeper.defaults :as zk-defaults]
            [franzy.admin.zookeeper.client :as client]
            [franzy.admin.cluster :as cluster]
            [clojure.java.io :as io]
            [clojure.string :as cstr]
            [taoensso.timbre :as log]
            [clojure.data.json :as json]
            [aero.core :as aero]
            [environ.core :refer [env]])
  (:import [org.apache.kafka.streams.kstream KStreamBuilder Predicate
ValueMapper]
            [org.apache.kafka.streams KafkaStreams StreamsConfig]
            [org.apache.kafka.common.serialization Serdes])
  (:gen-class))

(def api-endpoint "http://localhost:3000/api/")

(defn config [profile]
  (aero/read-config (io/resource "config.edn") {:profile profile}))

(defn get-broker-list
  [zk-conf]
  (let [c (merge (zk-defaults/zk-client-defaults) zk-conf)]
    (with-open[u (client/make-zk-utils c false)]
      (cluster/all-brokers u))))

(defn broker-str [zkconf]
  (let [zk-brokers (get-broker-list zkconf)
        brokers (map (fn [broker] (str (get-in broker [:endpoints :
plaintext :host]) ":" (get-in broker [:endpoints :plaintext :port]))  )
zk-brokers)]
    (if (= 1 (count brokers))
      (first brokers)
      (cstr/join "," brokers))))
```

```clojure
;; Kafka messages are still byte arrays at this point. Convert them to
strings.
(defn deserialize-message [bytes]
  (try (-> bytes
           java.io.ByteArrayInputStream.
           io/reader
           slurp)
       (catch Exception e (log/info (.printStackTrace e)))
       (finally (log/info ""))))

;; Function takes the byte array message and converts it to a Clojure
map.
(defn pre-process-data [data-in]
  (log/info "Pre process data")
  (log/info data-in)
  (let [message (-> data-in
                    deserialize-message
                    )
        json-out (json/read-str message :key-fn keyword)]
    (log/info json-out)
    json-out))

;; Process any commands, basically fire them at the HTTP API.
(defn process-command [data-in]
  (let [jsonm (pre-process-data data-in)]
    (slurp (str api-endpoint (:payload jsonm)))))

;; Extract the CSV training data and return it.
(defn process-training [data-in]
  (let [jsonm (pre-process-data data-in)]
    (.getBytes (:payload jsonm))))

;; Test the type key against the message, returns true/false
(defn pred-key-type [key message]
  (log/infof "Checking %s for key %s" message key)
  (let [b (if (= key (:type message))
            true
            false)]
    (log/infof "key %s - result is %b" key b)
    b))

;; This is the actual Kafka streaming application.
;; All the config is read in and then the app will figure out the rest.
;; The stream is branched to process the event stream (either a command
or training data)
(defn start-stream []
  (let [{:keys [kafka zookeeper] :as configuration} (config (keyword
(env :profile)))
```

```clojure
      _ (log/info "PROFILE" (env :profile))
      broker-list (broker-str {:servers zookeeper})
      props {StreamsConfig/APPLICATION_ID_CONFIG,  (:consumer-group
kafka)
             StreamsConfig/BOOTSTRAP_SERVERS_CONFIG, broker-list
             StreamsConfig/ZOOKEEPER_CONNECT_CONFIG, zookeeper
             StreamsConfig/TIMESTAMP_EXTRACTOR_CLASS_CONFIG
"org.apache.kafka.streams.processor.WallclockTimestampExtractor"
             StreamsConfig/KEY_SERDE_CLASS_CONFIG,   (.getName
(.getClass (Serdes/String)))
             StreamsConfig/VALUE_SERDE_CLASS_CONFIG, (.getName
(.getClass (Serdes/ByteArray)))}
      builder (KStreamBuilder.)
      config (StreamsConfig. props)
      input-topic (into-array String [(:topic kafka)])
      training-topic-name (:training-data kafka)
      training-topic (into-array String [training-topic-name])]
  (log/infof "Zookeeper Address: %s" zookeeper)
  (log/infof "Broker List: %s" broker-list)
  (log/infof "Kafka Topic: %s" (:topic kafka))
  (log/infof "Kafka Consumer Group: %s" (:consumer-group kafka))
  (do
    (let [partitioned-stream
          (.branch
           (.stream builder input-topic)
           (into-array Predicate [(reify
                                    Predicate
                                    (test [_ _ v]
                                      (do
                                        (log/info "p0 " v)
                                        (->> v
                                             (pre-process-data)
                                             (pred-key-type
"command")))))
                                  (reify
                                    Predicate
                                    (test [_ _ v]
                                      (do
                                        (log/info "p1 " v)
                                        (->> v
                                             (pre-process-data)
                                             (pred-key-type
"training")))))]))
          training-topic-stream (.stream builder training-topic)]
      (log/info partitioned-stream)
      (log/info training-topic-stream)
      (-> (aget partitioned-stream 0)
```

```
          (.mapValues (reify ValueMapper (apply [_ v] (process-command
  v))))
          (.print))
      (-> (aget partitioned-stream 1)
          (.mapValues (reify ValueMapper (apply [_ v] (process-
  training v))))
          (.to training-topic-name)))
    (KafkaStreams. builder config))))
```

Making Predictions

With Kafka handling the incoming events and training data, it's now time to turn your attention to making predictions against the built models. In the event topic streaming application, you may have noticed a reference to the API endpoint.

```
(def api-endpoint "http://localhost:3000/api/")
```

The code for the API is covered earlier in the chapter. For making predictions, the three prediction endpoints are used. They all work in the same way.

```
/predict/<model to use>/<csv values>
```

The model used in this instance is either the simple linear regression model (slr), the decision tree (dtr), or the neural network (mlp). The CSV values are the three scores, and the model will output the fourth. So, for example, a call to the API to predict the scores 3, 4, and 5 to the decision tree would look like this:

```
http://localhost:3000/api/predict/dtr/3,4,5
```

For predictions being made through Kafka via prediction_request_topic, there needs to be an application that will read the payload and run the prediction against the API.

Prediction Streaming API

The streaming application reads the message from the topic and deserializes it to a JSON format. The configuration of the application is similar to that of the event processing application shown earlier.

The real work happens in the run-prediction function.

```
(defn run-prediction [data-in]
  (let [jsonm (pre-process-data data-in)
        prediction-json (p/make-prediction jsonm)]
    (->> prediction-json
        (.getBytes))))
```

The value from the `make-prediction` function is returned and stored in the `prediction-json` value; at this point, it's a JSON format message but must be converted into a byte array so it can be sent to the prediction-response topic. The following is the full code for the prediction streaming application:

```
(ns kafka.stream.prediction.core
  (:require [franzy.admin.zookeeper.defaults :as zk-defaults]
            [franzy.admin.zookeeper.client :as client]
            [franzy.admin.cluster :as cluster]
            [clojure.java.io :as io]
            [clojure.string :as cstr]
            [taoensso.timbre :as log]
            [clojure.data.json :as json]
            [aero.core :as aero]
            [environ.core :refer [env]]
            [kafka.stream.prediction.predict :as p])
  (:import [org.apache.kafka.streams.kstream KStreamBuilder ValueMapper]
           [org.apache.kafka.streams KafkaStreams StreamsConfig]
           [org.apache.kafka.common.serialization Serdes])
  (:gen-class))

(def api-endpoint "http://localhost:3000/api/")

(defn config [profile]
  (aero/read-config (io/resource "config.edn") {:profile profile}))

(defn get-broker-list
  [zk-conf]
  (let [c (merge (zk-defaults/zk-client-defaults) zk-conf)]
    (with-open [u (client/make-zk-utils c false)]
      (cluster/all-brokers u))))

(defn broker-str [zkconf]
  (let [zk-brokers (get-broker-list zkconf)
        brokers (map (fn [broker] (str (get-in broker [:endpoints :
plaintext :host]) ":" (get-in broker [:endpoints :plaintext :port]))) )
zk-brokers)]
    (if (= 1 (count brokers))
      (first brokers)
      (cstr/join "," brokers))))

;; Kafka messages are still byte arrays at this point. Convert them to
strings.
(defn deserialize-message [bytes]
  (try (-> bytes
           java.io.ByteArrayInputStream.
           io/reader
           slurp)
       (catch Exception e (log/info (.printStackTrace e)))
       (finally (log/info "")))))
```

```clojure
;; Function takes the byte array message and converts it to a Clojure
map.
(defn pre-process-data [data-in]
  (log/info "Pre process data")
  (log/info data-in)
  (let [message (-> data-in
                    deserialize-message
                    )
        json-out (json/read-str message :key-fn keyword)]
    (log/info json-out)
    json-out))


;; Process any commands, basically fire them at the HTTP API.
(defn run-prediction [data-in]
  (let [jsonm (pre-process-data data-in)
        prediction-json (p/make-prediction jsonm)]
    (->> prediction-json
         (.getBytes))))


;; This is the actual Kafka streaming application.
;; All the config is read in and then the app will figure out the rest.
(defn start-stream []
  (let [{:keys [kafka zookeeper] :as configuration} (config (keyword
(env :profile)))
        _ (log/info "PROFILE" (env :profile))
        broker-list (broker-str {:servers zookeeper})
        props {StreamsConfig/APPLICATION_ID_CONFIG,   (:consumer-group
kafka)
               StreamsConfig/BOOTSTRAP_SERVERS_CONFIG, broker-list
               StreamsConfig/ZOOKEEPER_CONNECT_CONFIG, zookeeper
               StreamsConfig/TIMESTAMP_EXTRACTOR_CLASS_CONFIG
"org.apache.kafka.streams.processor.WallclockTimestampExtractor"
               StreamsConfig/KEY_SERDE_CLASS_CONFIG,   (.getName
(.getClass (Serdes/String)))
               StreamsConfig/VALUE_SERDE_CLASS_CONFIG, (.getName
(.getClass (Serdes/ByteArray)))}
        builder (KStreamBuilder.)
        config (StreamsConfig. props)
        input-topic (into-array String [(:input-topic kafka)])
        response-topic-name (:output-topic kafka)]
    (log/infof "Zookeeper Address: %s" zookeeper)
    (log/infof "Broker List: %s" broker-list)
    (log/infof "Kafka Topic: %s" (:input-topic kafka))
    (log/infof "Kafka Consumer Group: %s" (:consumer-group kafka))
    (do
      (->
        (.stream builder input-topic)
        (.mapValues (reify ValueMapper (apply [_ v] (run-prediction v))))
        (.to response-topic-name)))
    (KafkaStreams. builder config)))
```

Prediction Functions

When the streaming application calls the `make-prediction` function, it's referring to the predict namespace. Clojure has a mechanism for this called *multi-methods*, which enable you to create a single function name but have various functions complete the work.

Multi-methods are perfect for the prediction mechanism as we're passing in the JSON payload and dealing with two aspects, the model type and the values to predict against.

Each function has the same structure:

```
(defmethod make-prediction :my-model-type [event]
  (json/write-str (my-model-type-function/predict-model (:payload
event))))
```

It's the same function throughout; it's just the model name that changes. This makes our application extendable easily. Perhaps I decide to add a support vector machine model to the system. All I have to do is add a new multi-method to call the prediction on the support vector machine code I've written. In the predict namespace, I'd add the following:

```
(defmethod make-prediction :svm [event]
  (json/write-str (svm/predict-model (:payload event))))
```

And that's it—there's no refactoring of the streaming application required. For now, there are three models in the project, and the code listing for the prediction calls looks like this:

```
(ns kafka.stream.prediction.predict
  (:require [kafka.stream.prediction.decisiontree :as dt]
            [kafka.stream.prediction.linear :as lr]
            [kafka.stream.prediction.mlp :as mlp]
            [clojure.data.json :as json]))

(defn get-model-type [event]
  (keyword (:model event)))

(defmulti make-prediction (fn [event] (get-model-type event)))

(defmethod make-prediction :mlp [event]
  (json/write-str (mlp/predict-mlp (:payload event))))

(defmethod make-prediction :slr [event]
  (json/write-str (lr/predict-simple-linear (:payload event))))

(defmethod make-prediction :dtr [event]
  (json/write-str (dt/predict-decision-tree (:payload event))))
```

The last thing to look at is how the three models make predictions; they all do predictions differently as they all use different Java APIs. First is the decision tree.

Predicting with Decision Tree Models

The decision tree model uses Weka, and this namespace handles both the model build (via a shell script) and the predictions.

When the prediction mechanism calls a :dtr event, it executes the predict-decision-tree function. This does several tasks:

- It creates an ARFF format instance.
- It performs a query against the MySQL database to find the most accurate decision tree model, and the query returns the UUID.
- With the UUID, it then loads the model into memory.
- The prediction is made against the model.
- A JSON payload is created and returned to the calling function from the streaming application.

```
(ns prediction.http.api.decisiontree
  (:require [prediction.http.api.db :as db]
            [clj-time.core :as t]
            [clj-time.format :as f])
  (:use [clojure.java.shell :only [sh]])
  (:import [java.io ByteArrayInputStream InputStream InputStreamReader
BufferedReader]
           [weka.core Instances SerializationHelper]))

(def script-path "/opt/mlbook/work/strata-2018-kafka-dl4j-clojure/
projects/dl4j.mlp/scripts/rundtr.sh")

(defn run-model-build []
  (sh script-path)
  "Decision Tree built.")

(defn get-most-accurate-model []
  (first (db/load-accurate-model-by-type {:model-type "dtr"})))

(defn create-instance [input]
  (let [header (slurp "/opt/mlbook/testdata/wekaheader.txt")]
    (->> (str header input ",?")
         .getBytes
         (ByteArrayInputStream.)
         (InputStreamReader.)
         (BufferedReader.)
         (Instances.)) ))
```

```
(defn load-model [uuid]
  (let [model-path (str "/opt/mlbook/testdata/models/" uuid ".model")]
    (SerializationHelper/read model-path)))

(defn classify-instance [model instance]
  (do
    (.setClassIndex instance(- (.numAttributes instance) 1))
    (.classifyInstance model (.instance instance 0))))

(defn predict-decision-tree [x]
  (let [instance (create-instance x)
        model-info (get-most-accurate-model)
        model (load-model (:uuid model-info))
        result (classify-instance model instance)]
    {:input x
     :result result
     :accuracy (:model_accuracy model-info)
     :modelid (:uuid model)
     :prediction-date (f/unparse (f/formatters :mysql) (t/now))}))
```

Predicting Linear Regression

Both the decision tree and neural network models have serialized model builds persisted to disk. With the linear regression model, the slope and the intercept are stored within the MySQL database and therefore, don't need to deserialize anything; all that is required is a query to find the model with the highest R2 score.

This model does not use all four values to get a prediction, so there is a function to parse out the first value and convert it to an integer. This is what the prediction will be made against.

Notice that the majority of the code is parsing and converting. The actual work is done in two lines of code.

```
(defn calc-linear [slope intercept x]
  (+ intercept (* x slope)))
```

One of the beauties of the Clojure language is that you can achieve a high level of functionality with some concise code. The full code listing for the linear regression builder and predictor is shown here:

```
(ns prediction.http.api.linear
  (:require [prediction.http.api.db :as db]
            [clojure.string :as s]
            [clj-time.core :as t]
            [clj-time.format :as f])
  (:use [clojure.java.shell :only [sh]]))
```

```clojure
(def script-path "/opt/mlbook/work/strata-2018-kafka-dl4j-clojure/
projects/dl4j.mlp/scripts/runslr.sh")

(defn run-model-build []
  (sh script-path)
  "Linear model built.")

(defn load-simple-linear []
  (db/load-linear-model))

(defn calc-linear [slope intercept x]
  (+ intercept (* x slope)))

(defn convert-input-to-integer [input]
  (-> input
      (s/split #",")
      first
      (Integer/parseInt)))

;; load highest r2 valued model
(defn predict-simple-linear [x]
  (let [model (first (load-simple-linear))
        input (convert-input-to-integer x)
        result (calc-linear (:slope model)
                            (:intercept model)
                            input)]
    {:input input
     :result result
     :accuracy (:rsq model)
     :modelid (:uuid model)
     :prediction-date (f/unparse (f/formatters :mysql) (t/now))}))
```

Predicting the Neural Network Model

Similar to the decision tree model, the system has to do a number of steps before it gets to its prediction.

1. Find the most accurate model from the database, returning the UUID of the model.
2. Load the model into memory.
3. Parse the input query string from the API.
4. Run the prediction.
5. Build a JSON payload to return to the API as a response.

The full code listing takes care of the model build and the prediction. Building the model is a case of triggering the shell script in the project.

```clojure
(ns prediction.http.api.mlp
  (:require [prediction.http.api.db :as db]
            [clojure.string :as s]
            [clj-time.core :as t]
            [clj-time.format :as f])
  (:use [clojure.java.shell :only [sh]])
  (:import [org.deeplearning4j.util ModelSerializer]
           [org.nd4j.linalg.factory Nd4j]))

(def script-path "/opt/mlbook/work/strata-2018-kafka-dl4j-clojure/
projects/dl4j.mlp/scripts/runmlp.sh")

(defn run-model-build []
  (sh script-path)
  "Neural Network built.")

(defn get-most-accurate-model []
  (first (db/load-accurate-model-by-type {:model-type "mlp"})))

(defn build-model-filepath [uuid]
  (str "/opt/mlbook/testdata/models/" uuid ".zip"))

(defn load-mlp-model [uuid]
  (ModelSerializer/restoreMultiLayerNetwork (build-model-filepath
uuid)))

(defn split-input [input]
  (double-array
   (map #(Double/parseDouble %)
        (-> input
            (s/split #",")))))

(defn make-prediction [model input]
  (let [input-vector (Nd4j/create (split-input input))
        prediction (.output model input-vector)]
    (.iamax (Nd4j/getBlasWrapper) prediction)))

(defn predict-mlp [x]
  (let [model-info (get-most-accurate-model)
        model (load-mlp-model (:uuid model-info))
        prediction (make-prediction model x)]
    {:input x
     :result prediction
     :accuracy (:model_accuracy model-info)
     :modelid (:uuid model-info)
     :prediction-date (f/unparse (f/formatters :mysql) (t/now))}))
```

Running the Project

This is a big project with a lot of services that are running. Here's a quick run-down of how to get it all running. I'm assuming you are in the root directory of the project.

Run MySQL

Assuming MySQL is already running, create the tables.

```
$ mysqladmin -u root -p<your admin password> create mlchapter12
$ mysql -u root -p<your admin password> mlchapter12 < schema.sql
```

Run Zookeeper

Open a new terminal window and run the following command as the root user:

```
$ /path/to/kafka/bin/zookeeper-server-start.sh /path/to/kafka/config/
zookeeper.properties
```

Run Kafka

Open a new terminal window and run the following command as the root user:

```
$ /path/to/kafka/bin/kafka-server-start.sh /path/to/kafka/config/server.
properties
```

Create the Topics

Create the topics (in the `scripts` directory).

```
$ /path/to/project/scripts/create-topics.sh
```

Run Kafka Connect

Run as the root user and change the directory names to reflect your directory names.

```
$ /path/to/kafka/bin/connect-standalone.sh /path/to/kafka/config/
connect-standalone.properties
/path/to/project/config/dl4j_event_to_fs_sink.properties /path/to/
project/config/dl4j_to_fs_sink.properties
```

Model Builds

Go to the `dl4j.mlp` project in the `projects` folder and run the following in a new terminal window:

```
$ mvn package
```

Run Events Streaming Application

Go to the `kafka.stream.events` project in the `projects` folder and run the following command:

```
$ lein uberjar
$ java -jar target/uberjar/kafka-stream-events.jar
```

Run Prediction Streaming Application

Go to the `kafka.stream.prediction` project in the `projects` folder and run the following command:

```
$ lein uberjar
$ java -jar target/uberjar/kafka-stream-prediction.jar
```

Start the API

Go to the `prediction.http.api` project in the `projects` folder and run the following command:

```
$ lein uberjar
$ java -jar target/server.jar
```

Send JSON Training Data

With Zookeeper, Kafka, and Kafka Connect running, you can now send some data to `event_topic`. The following script will iterate each line of the JSON file in the data directory and pipe it to a Kafka console producer:

```
for i in `cat data/trainingdata.json` ; do echo $i ;done | /path/to/
kafka/bin/kafka-console-producer.sh --broker-list localhost:9092 --topic
event_topic
```

Train a Model

In the `messages` directory of the project you'll find three payloads to send a command event to `event_topic`. To request a build of the simple linear regression model, you can run the following command from the command line:

```
cat build_slr.json | /path/to/kafka/bin/kafka-console-producer.sh
--broker-list localhost:9092 --topic event_topic
```

Make a Prediction

There is a sample prediction message in the `messages` directory. In the same way you requested a model build, `sample_predict.json` to `prediction_request_topic`.

```
cat sample_predict.json | /path/to/kafka/bin/kafka-console-producer.sh
--broker-list localhost:9092 --topic prediction_request_topic
```

If you are running a consumer on the `prediction_response_topic`, you will see the JSON output with the prediction.

```
{"input": "3,4,5"
 "result": 5.27
 "accuracy":76.876
 "modelid" "32201ab39be3745e5e9a7e576827cc59"
 "predicton-date": "2019-08-11 18:58:00"}
```

Summary

Kafka provides us with a solid system to produce and consume messages in near real time. By combining it with other technologies from your acquired machine learning knowledge you can build a system that uses streaming data to create a set of models to make predictions, over time improving on the accuracy of the models as new data is added.

While the components of the Kafka framework are simple the potential to create streaming intelligence applications are huge. It's about knowing how to connect each element together and refining the process. This chapter covered many aspects of data collection, processing, training, and prediction.

Apache Spark

The Apache Spark project was created by the AMPLab at UC Berkeley as a data analytics cluster computing framework. This chapter is a quick overview of the Scala language and its use within the Spark framework. The chapter also looks at the external libraries for machine learning, SQL-like queries, and streaming data with Spark.

Spark: A Hadoop Replacement?

The debate about whether Spark is a Hadoop replacement might rage on longer than some would like. One of the problems with Hadoop is the same thing that made it famous: MapReduce. The programming model can take time to master for certain tasks. If it's a case of straight totaling up frequencies of data, then MapReduce is fine, but after you get past that point, you're left with some hard decisions to make.

Hadoop2 gets beyond the issue of using Hadoop only for MapReduce. With the introduction of YARN (Yet Another Resource Negotiator), Hadoop acts as an operating system for data with YARN controlling resources against the cluster. These resources weren't limited to MapReduce jobs; they could be any job that could be executed.

The Spark project doesn't rely on MapReduce, which gives it a speed advantage. The claim is that it's 100 times faster than in-memory Hadoop and 10 times faster on disk. If speed is an issue for you, then Spark is certainly up there on the list of things to look at. As for the argument that it's a replacement for Hadoop, well, there's a time and place to use Hadoop, and the same goes for Spark; it's all about the project that you are completing and picking the right tools for the job. The tools are just that—tools.

Java, Scala, or Python?

Spark jobs can be written in Java, Scala, or Python. As usual, which language you use tends to be a matter of personal preference.

The book has so far concentrated on Java as its core language of choice, and I will continue using the Java language for the Spark examples. During the first edition of this book, the Python libraries for Spark were classed as experimental; these have been updated and more widely used as Python, as a language, was more widely adopted for data science.

As you progress through this chapter, you might come to the conclusion that Java is too bulky for quickly writing Spark jobs. Remember, it's a matter of personal preference, so you should use the language in which you're most comfortable.

Downloading and Installing Spark

There are a few ways to download and use Spark. The easiest way is to use the prebuilt packages that are available from the Spark website. Before you download one, check which version of Hadoop you (or your organization) is running as it will determine the download you want. If you are still running Hadoop, then it's preferable to match your Spark download to the core Hadoop version you're running. For most developers, I would wager this is no longer an issue.

After you have downloaded the file for your Hadoop version, you can install it by moving the downloaded file to the directory you want to install it to. The file is a .tgz file, so you can unarchive it in one command.

```
tar xvzf spark-2.4.4-bin-hadoop2.7.tgz
```

The contents of the file will unarchive and be ready for use.

A Quick Intro to Spark

The interactive shell in Scala gives you a quick and easy way to see what Spark can do in a short span of time. Don't worry if you haven't used Scala before; the examples shown here are simple. It's handy to know a few of the basic Spark lines in Scala so you can inspect data from the REPL and not have to create code to accomplish the task. I'll pick up on how to do full applications in Spark after this short introduction.

Starting the Shell

From the directory where you installed Spark, type the following command to launch the shell:

```
./bin/spark-shell
```

You see the following output while Spark boots up:

```
$ bin/spark-shell
19/10/20 09:24:25 WARN NativeCodeLoader: Unable to load native-hadoop
 library for your platform... using builtin-java classes where
applicable
Using Spark's default log4j profile: org/apache/spark/log4j-
defaults.properties
Setting default log level to "WARN".
To adjust logging level use sc.setLogLevel(newLevel). For SparkR,
use setLogLevel(newLevel).
Spark context Web UI available at http://192.168.1.102:4040
Spark context available as 'sc' (master = local[*],
app id = local-1571559873516).
Spark session available as 'spark'.
Welcome to
      ____              __
     / __/__  ___ _____/ /__
    _\ \/ _ \/ _ `/ __/  '_/
   /___/ .__/\_,_/_/ /_/\_\   version 2.4.4
      /_/

Using Scala version 2.11.12 (Java HotSpot(TM) 64-Bit Server VM,
Java 1.8.0_45)
Type in expressions to have them evaluated.
Type :help for more information.

scala>
```

I'm showing all the messages, because there are a few interesting things I want to show you in a moment. For now, though, you should see the `scala>` prompt at the bottom, which means you're ready.

Data Sources

Spark supports the same input file systems as Hadoop. If your data store is supported by the `InputFormat` method in Hadoop, then you can read it into Spark without too much effort.

The obvious ones that come to mind are the local filesystem, Amazon S3 buckets, HBase, Cassandra, and files already on a Hadoop Distributed File System (HDFS). As you see in the next section, you're using the `sc.textFile` method to load in data. This isn't limited to specific files; you can use that method to process wildcards on files, zipped files, and directories as well.

Testing Spark

Find a text file with which to test Spark and follow along with the rest of this section. I'm using a file from the local filesystem for this example.

Load the Text File

First, load the text file. From the Scala command line, type the following:

```
scala> var textF = sc.textFile("/path/to/data/ch13/mobydick.txt")
```

Basically you're storing the contents of the text file into a Scala variable called `textF`. Spark responds with output along the lines of the following:

```
textF: org.apache.spark.rdd.RDD[String] =
/path/to/data/ch13/mobydick.txt MapPartitionsRDD[3] at textFile at
<console>:24
```

Spark uses a concept called *resilient distributed datasets* (RDDs), so in the output you can see that you now have a MappedRDD containing strings. With the text file loaded, you can start to inspect it and get some results.

Make Some Quick Inspections

With the data loaded, you can do some quick inspections. First, how many elements of data do you have in the RDD?

```
scala> textF.count()
```
You get the following output:

```
res1: Long = 7182
scala>
```

This count represents the number of elements in the RDD and not the number of lines in the text file. You can pull the first element from the RDD, as shown here:

```
scala> textF.first()
res2: String = CHAPTER 1
scala>
```

As you can see from the output, these results are returning quickly as the RDD is based in memory.

Filter Text from the RDD

With the `.filter` function, you can start to inspect specific things within the text file. Assuming that you want to see how many times the word *statistical* occurs in the document, you can run the following:

```
scala> textF.filter(line => line.contains("whale")).count()
```

One line in Scala and the Spark filter iterate the RDD and inspect the lines. Because you've appended the count at the end, you get the following result:

```
res3: Long = 316
```

So, there are 316 mentions of the word *whale* in the entire document. If required, you could save this output to another Scala array.

```
scala> var filtered = textF.filter(line => line.contains("whale"))
filtered: org.apache.spark.rdd.RDD[String] = MapPartitionsRDD[6]
at filter at <console>:25

scala> filtered.count()
res4: Long = 316
```

With the aid of concise commands in Scala and the power of in-memory processing of distributed datasets (the RDDs), you have a neat system to get large amounts of crunching done quickly.

Spark Monitor

Earlier, I mentioned that when you start Spark, a few things are being run. One of them is the web-based monitor. If you point your browser to `http://<yourdomain>:4040`, you should get the website shown in Figure 13.1, assuming Spark is still running.

As far as Spark is concerned, every line you run is a job, so Spark logs it accordingly, giving its duration and outcome. There's also information on the storage and runtime environment.

You can click each of the stages to see full details of the job and its execution output. If one of your jobs is causing trouble, then it's handy to look here first and get a bird's-eye view of things.

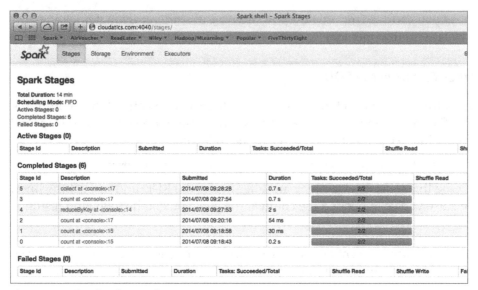

Figure 13.1: Spark web console

Comparing Hadoop MapReduce to Spark

So, how does a basic Spark job compare to the same Hadoop job? Let's look at an example. Hadoop jobs usually are built around the MapReduce paradigm that contains, not surprisingly, a map phase and a reduce phase.

The basic Java code boilerplate is usually along the lines of the following:

```
public static class Map extends MapReduceBase implements
        Mapper<LongWritable, Text, Text, IntWritable> {
    private final static IntWritable one = new IntWritable(1);
    private Text word = new Text();

    public void map(LongWritable key, Text value,
            OutputCollector<Text, IntWritable> output,
Reporter reporter)
            throws IOException {
        // ususally emit something to the reducer here....
    }
}

    public static class Reduce extends MapReduceBase implements
            Reducer<Text, IntWritable, Text, IntWritable> {
        public void reduce(Text key, Iterator<IntWritable> values,
            OutputCollector<Text, IntWritable> output,
```

```
Reporter reporter)
            throws IOException {
        // reducer would add the value +1 for example


    }
  }
```

Then a job definition enables it to run within the Hadoop framework; additional information is added like the input and output formats and the paths to use to read the input data and where to write the results.

```
public static void main(String[] args) throws IOException {
        JobConf conf = new JobConf(BlankHadoopJob.class);
        conf.setJobName("BlankHadoopJob");

        conf.setOutputKeyClass(Text.class);
        conf.setOutputValueClass(IntWritable.class);

        conf.setMapperClass(Map.class);
        conf.setCombinerClass(Reduce.class);
        conf.setReducerClass(Reduce.class);

        conf.setInputFormat(TextInputFormat.class);
        conf.setOutputFormat(TextOutputFormat.class);

        FileInputFormat.setInputPaths(conf, new Path(args[0]));
        FileOutputFormat.setOutputPath(conf, new Path(args[1]));

        JobClient.runJob(conf);


    }
```

Some developers complain about the amount of code you need to write in Java to get MapReduce working. It's never been an issue for me (as I have templates set up), but when I show you how it works in Spark, you'll realize why they were complaining.

In Spark, you can put together a quick word count MapReduce routine that demonstrates how easy it is to do. Using the same text file you used earlier, you can run a MapReduce process from the Spark shell. First load the text file.

```
scala> var textF = sc.textFile("/path/to/data/ch13/mobydick.txt")
```

Then create a new variable with the results of the MapReduce.

```
scala> var mapred = textF.flatMap(line => line.split(" ")).
map(word => (word, 1)).reduceByKey((a,b) => a+b)
mapred: org.apache.spark.rdd.RDD[(String, Int)] = ShuffledRDD[9]
at reduceByKey at <console>:25
```

Then output the results.

```
scala> mapred.collect
```

You see a block of the output appear in the shell:

```
res5: Array[(String, Int)] = Array((shelf,,2), (Ah!,5), (bone,6),
(lug,1), (roses.,1), (dreamiest,,1), (dotings,1), (Fat-Cutter;,1),
(seems--aye,,1), (Boat,1), (countrymen.,1), (consideration,,1),
(chapters,3), (sweat.,1), (pants.,1), (wasn't,3), (been,116), (they,,2),
(chests,,1), (proceedings;,1), (battering,2), (contemptible,2),
(salt-sea,2), (knows,8), (fowl,1), (inch.,1), (Nathan,1), (surf.,1),
(ignore,1), (greenness,,1), (angels,,1), (smooth,1), (stern,5),
(casket,--the,1), (completing,1), (seductive,1), (disdain,,1),
(disclosures,,1), (snuffing,1), (southward;,1), (Steady,,1),
(erected,3), (hypocrisies,1), (dead,20), (savages,,1), (eloquently,1),
(Pots.,1), (thee,,6), (regardful,1), (startling,2), (thus,15),
(Huzza,4), (historians,1), (descending,3), (crowned,1), (iron,13),
(seve...
```

To save the results, use the `.saveAsTextFile` method on the RDD to output as text (change the output directory to match your home directory, for example).

```
scala> mapred.saveAsTextFile("/path/to/output/mapred_testoutput")
```

Spark, in the same way as Hadoop, saves the files in a directory (I called this one `testoutput`). Within it you see the `part-00000` files.

```
-rw-r--r-- 1 1234 1234 52961 Jul  8 13:47 part-00000
-rw-r--r-- 1 1234 1234 52861 Jul  8 13:47 part-00001
-rw-r--r-- 1 1234 1234     0 Jul  8 13:47 _SUCCESS
```

The output of those files contains the basic word count.

```
$ less /path/to/output/mapred_testoutput/part-00000
(shelf,,2)
(Ah!,5)
(bone,6)
(lug,1)
(roses.,1)
(dreamiest,,1)
(dotings,1)
(Fat-Cutter;,1)
(seems--aye,,1)
(Boat,1)
(countrymen.,1)
(consideration,,1)
(chapters,3)
(sweat.,1)
(pants.,1)
(wasn't,3)
```

```
(been,116)
(they,,2)
(chests,,1)
(proceedings;,1)
(battering,2)
(contemptible,2)
```

In three lines you performed a basic MapReduce program on some raw text. Notice that I didn't remove odd characters and convert everything to lowercase, but essentially it gave us the word count output.

Writing Stand-Alone Programs with Spark

While it is possible to perform some basic transforms and applications of scripts with the Spark REPL, the real power comes from the APIs that are available. This means that it's possible to build big data applications in Java, Python, Scala, or any other supported language. For the duration of this chapter, I will keep with Java. If you are a Clojure developer, then it's worth investigating the Sparkling Spark wrappers for Clojure (`http://gorillalabs.github.io/sparkling/`).

Spark Programs in Java

Earlier in the chapter I showed you how to perform a basic word count on the Moby Dick text. For this example, I'll cover the basics of writing an application in Java to do the same in Spark. The workflow is broken down into three parts.

- The code
- The Maven build file (`pom.xml`)
- Deployment to Spark to run the application

The following Java code is the basic word count application:

```
package mlbook.ch13.spark;

import org.apache.spark.api.java.JavaRDD;
import org.apache.spark.sql.SparkSession;
import scala.Tuple2;
import org.apache.spark.api.java.JavaPairRDD;
import java.util.Arrays;
import java.util.List;
import java.util.regex.Pattern;

public final class BasicSparkWordCount {
    private static final Pattern SPACE = Pattern.compile(" ");
```

```
public static void main(String[] args) throws Exception {
    System.out.println("Starting BasicSparkWordCount....");
    if (args.length < 1) {
        System.err.println("Usage: BasicSparkWordCount <file>");
        System.exit(1);
    }

    SparkSession sparkSession = SparkSession
            .builder()
            .appName("BasicSparkWordCount")
            .getOrCreate();

    JavaRDD<String> linesOfText = sparkSession.read()
.textFile(args[0]).javaRDD();
    JavaRDD<String> wordsInEachLine = linesOfText
.flatMap(s -> Arrays.asList(SPACE.split(s)).iterator());
    JavaPairRDD<String, Integer> allTheOnes = wordsInEachLine
.mapToPair(singleWord -> new Tuple2<>(singleWord, 1));
    JavaPairRDD<String, Integer> finalCounts = allTheOnes
.reduceByKey((i1, i2) -> i1 + i2);

    List<Tuple2<String, Integer>> output = finalCounts.collect();
    for (Tuple2<?, ?> tuple : output) {
        System.out.println("Segment " + tuple._1() + " found "
+ tuple._2() + " times.");
    }
    sparkSession.stop();
    System.out.println("Finishing BasicSparkWordCount....");
    }
}
```

The process flow of the application works as follows. The `SparkSession` handles all the setup for the Spark application. Any specific config settings and naming would happen at this point. The following four lines are where the real work is happening.

The first stage is using the `sparkSession.read()` function to load the data into RDDs. These are just blocks of strings from the text file. The second stage is to split the words in each line of text by the space character. The resulting transform is an array of words.

Now we're at the third stage, where the `mapToPair` function generates a new RDD with the word as the key and the number 1 as a value. The final RDD reduces each instance of the word/value pairs in the `allTheOnes` RDD.

To output the results, the `.collect` function is used. This applies the Spark job in the session and collects the reduced results from the `finalCounts` RDD. The resulting list is iterated, and the results are displayed to the standard output.

Using Maven to Build the Project

Maven is now used as the build tool of choice for the majority of Java applications. If you don't have Maven installed, you can download it from http://maven.apache.org and unarchive the file.

For every project, you need a Maven build file; this is called pom.xml. For this application, the build file looks like this:

```xml
<?xml version="1.0" encoding="UTF-8"?>
<project xmlns="http://maven.apache.org/POM/4.0.0"
        xmlns:xsi="http://www.w3.org/2001/XMLSchema-instance"
        xsi:schemaLocation="http://maven.apache.org/POM/4.0.0
 http://maven.apache.org/xsd/maven-4.0.0.xsd">
   <modelVersion>4.0.0</modelVersion>
   <groupId>mlbook</groupId>
   <artifactId>Chapter13</artifactId>
   <version>1.0.0-SNAPSHOT</version>

   <properties>
       <sbt.project.name>sparkChapter13</sbt.project.name>
       <java.version>1.8</java.version>
       <maven.compiler.source>1.8</maven.compiler.source>
       <maven.compiler.target>1.8</maven.compiler.target>
       <build.testJarPhase>none</build.testJarPhase>
       <build.copyDependenciesPhase>package
</build.copyDependenciesPhase>
   </properties>

   <dependencies>
       <dependency>
           <groupId>org.apache.spark</groupId>
           <artifactId>spark-core_2.11</artifactId>
           <version>2.2.0</version>
       </dependency>
       <dependency>
           <groupId>org.apache.spark</groupId>
           <artifactId>spark-sql-kafka-0-10_2.11</artifactId>
           <version>2.2.0</version>
       </dependency>
       <dependency>
           <groupId>org.apache.spark</groupId>
           <artifactId>spark-sql_2.11</artifactId>
           <version>2.2.0</version>
       </dependency>
   </dependencies>
```

```
    <build>
        <outputDirectory>target/scala-2.11/classes</outputDirectory>
        <testOutputDirectory>target/scala-2.11/test-classes
</testOutputDirectory>
        <plugins>
            <plugin>
                <groupId>org.apache.maven.plugins</groupId>
                <artifactId>maven-deploy-plugin</artifactId>
                <configuration>
                    <skip>true</skip>
                </configuration>
            </plugin>
            <plugin>
                <groupId>org.apache.maven.plugins</groupId>
                <artifactId>maven-install-plugin</artifactId>
                <configuration>
                    <skip>true</skip>
                </configuration>
            </plugin>
            <plugin>
                <groupId>org.apache.maven.plugins</groupId>
                <artifactId>maven-jar-plugin</artifactId>
                <configuration>
                    <outputDirectory>${jars.target.dir}</outputDirectory>
                </configuration>
            </plugin>
        </plugins>
    </build>
</project>
```

This gives you the basic outline of the project and which repositories to pull any required dependencies from. For Spark projects, you need to have the Spark API in the dependency declaration. As this chapter continues, you will be adding more dependencies as you go along; the full pom build file is in the code repository that accompanies the book if you don't want to type in the entire build file.

Creating Packages in Maven

To create the package, you run Maven from the command line.

```
mvn package
```

The Maven build tool looks after the downloading of the dependencies, creates the class files, and then packages the .jar file with the required classes. After the package is built, a directory called target is created, and you see the .jar file there.

```
$ pwd
/path/to/code/java/ch13/target
```

```
$ ls -l
total 16
-rw-r--r--  1 jasebell staff  5917 22 Oct 09:49
Chapter13-1.0.0-SNAPSHOT.jar
drwxr-xr-x  3 jasebell staff    96 21 Oct 10:44 generated-sources
drwxr-xr-x  3 jasebell staff    96 21 Oct 10:44 maven-archiver
drwxr-xr-x  3 jasebell staff    96 21 Oct 10:42 maven-status
drwxr-xr-x  3 jasebell staff    96 21 Oct 11:25 scala-2.11
```

To run the project with Spark, you need to use the `spark-submit` program like you did in the previous Scala example (make sure it's one continuous line in your terminal window).

```
$ /usr/local/spark-2.4.4-bin-hadoop2.7/bin/spark-submit --class \
"mlbook.ch13.spark.BasicSparkWordCount" --master local[4] \
/path/to/Chapter13/target/Chapter13-1.0.0-SNAPSHOT.jar \
/path/to/data/ch13
```

Let's take a closer look at the command line; there are a few things to explain. First, there's the `--class` flag that tells Spark which class to run; it's using the full package name, and it's encased in quotes. The `--master` flag tells Spark which cluster to run the job on; in this instance, it's on the local machine and using four threads to run the job on. The last two parts are the location of the JAR file and then any other arguments for the executing application (in this case, the path to the text file directory).

Spark executes the `.jar` file, and you see the MapReduce output in the console, shown here:

```
Starting BasicSparkWordCount....
Segment Ah! found 5 times.
Segment Let found 11 times.
Segment lug found 1 times.
Segment roses. found 1 times.
Segment bone found 6 times.
Segment dreamiest, found 1 times.
Segment Fat-Cutter; found 1 times.
Segment seems--aye, found 1 times.
Segment Boat found 1 times.
Segment wasn't found 3 times.
Segment Imperial found 1 times.
Segment been found 116 times.
Segment end found 10 times.
Segment they, found 2 times.
```

While the Spark application has done exactly what was expected, it would be prudent as an exercise to tidy up the data before processing. Take a look at Chapter 3 for further information.

Spark Program Summary

With our first Spark application created, let's expand on this knowledge and look at some of the other APIs that Spark provides: SparkSQL, Spark Streaming, and MLLib.

You can build all the applications in the same way. By refactoring the `pom.xml` build file, you can quickly expand the Spark libraries to your applications.

Spark SQL

The Big Data world has moved on a lot since its Hadoop heyday. Sadly, no one really talks about Pig scripts anymore. The introduction of SparkSQL gave us a system to run high-performance queries against large datasets.

Since the first edition of this book, the Spark SQL system went through an overhaul that introduced data frames and a more robust way of handling queries. In this section, you will see how to build up a Java Spark application to load CSV data and run queries against it.

Basic Concepts

As you'll remember from the previous Spark application example, `SparkSession` creates the Spark environment that will run your application in the cluster. The SparkSQL libraries work in the same way.

To illustrate the different layers of the SparkSQL functionality, I'm going to start with a basic Spark job and add the SQL methods to do different things from the API.

```
package mlbook.ch13.spark;

import org.apache.spark.sql.AnalysisException;
import org.apache.spark.sql.Dataset;
import org.apache.spark.sql.Row;
import org.apache.spark.sql.SparkSession;

public class BasicSparkSQL {
    private static String airportDataPath =
"/path/to/data/ch13/sql/airports.csv";

    public static void main(String[] args) throws AnalysisException {
        SparkSession spark = SparkSession
                .builder()
                .appName("ML Book Spark SQL Example")
                .getOrCreate();
        // we'll add more here soon!
        spark.stop();
    }
}
```

I've added a string that contains the path to the data you will be querying against. It's a CSV file of airport data. From here, I'll start adding some methods to do the work.

Let's start by creating some output to make sure the data file is being loaded in properly; if problems are going to occur, I've found it's usually finding the data in the first place.

In your code, create a new method called `runShowAirports`.

```
public static void runShowAirports(SparkSession spark)
throws AnalysisException {
   Dataset<Row> df = spark.read().csv(airportDataPath);
   df.show();
}
```

Spark will read the CSV file into a data frame; there's no converting or iterating over the data to get the rows into RDDs. Add the method call in your `main()` method after the Spark session is created.

```
runShowAirports(spark);
```

The next step is to build and run the application. Like earlier, run `mvn package` to create a new JAR file in the same target location as the previous application. The next step is to execute it.

```
$ /usr/local/spark-2.4.4-bin-hadoop2.7/bin/spark-submit \
--class "mlbook.ch13.spark.BasicSparkSQL" --master local[4] target/
Chapter13-1.0.0-SNAPSHOT.jar
```

Notice that I've removed the reference to the data path; the other thing that has changed is the package and class name to execute. When you run the application, you'll see Spark start up, load the data, and dump the first 20 rows to the console.

```
19/10/22 14:20:58 INFO TaskSchedulerImpl: Removed TaskSet 1.0, whose
tasks have all completed, from pool
19/10/22 14:20:58 INFO DAGScheduler: ResultStage 1 (show at
BasicSparkSQL.java:27) finished in 0.139 s
19/10/22 14:20:58 INFO DAGScheduler: Job 1 finished: show at
BasicSparkSQL.java:27, took 0.143854 s
+---+-------------------+------------+----------------+----+----+----
-----+----------+----+--------+----+--------------------+
|_c0|                 _c1| _c2|             _c3| _c4| _c5| _c6| _c7| _c8|
      _c9|_c10| _c11|
+---+-------------------+------------+----------------+----+----+----
-----+----------+----+--------+----+--------------------+
| id|               name| city|         country|iata|icao| lat|
lon| alt|timezone| dst|        tz|
|  1|             Goroka| Goroka|Papua New Guinea| GKA|AYGA|-
6.081689|145.391881|5282|      10| U|Pacific/Port_Moresby|
|  2|             Madang| Madang|Papua New Guinea| MAG|AYMD|-5.207083|
145.7887| 20| 10| U|Pacific/Port_Moresby|
```

```
|   3|          Mount Hagen|   Mount Hagen|Papua New Guinea| HGU|AYMH|-
5.826789|144.295861|5388|     10| U|Pacific/Port_Moresby|
|   4|              Nadzab| Nadzab|Papua New Guinea|
LAE|AYNZ|-6.569828|146.726242| 239|     10| U|Pacific/Port_Moresby|
|   5|Port Moresby Jack...|   Port Moresby|Papua New Guinea|
POM|AYPY|-9.443383| 147.22005| 146|     10| U|Pacific/Port_Moresby|
|   6|          Wewak Intl|      Wewak|Papua New Guinea|
WWK|AYWK|-3.583828|143.669186|  19| 10| U|Pacific/Port_Moresby|
|   7|           Narsarsuaq| Narssarssuaq|       Greenland|
UAK|BGBW|61.160517|-45.425978| 112|     -3| E| America/Godthab|
|   8|                Nuuk| Godthaab|       Greenland|
GOH|BGGH|64.190922|-51.678064| 283|     -3| E| America/Godthab|
|   9|   Sondre Stromfjord|   Sondrestrom| Greenland|
SFJ|BGSF|67.016969|-50.689325| 165|     -3| E| America/Godthab|
|  10|     Thule Air Base|        Thule| Greenland|
THU|BGTL|76.531203|-68.703161| 251|     -4| E| America/Thule|
|  11|            Akureyri| Akureyri|       Iceland|
AEY|BIAR|65.659994|-18.072703|   6| 0| N| Atlantic/Reykjavik|
|  12|         Egilsstadir| Egilsstadir|       Iceland|
EGS|BIEG|65.283333|-14.401389|  76| 0| N| Atlantic/Reykjavik|
|  13|         Hornafjordur|        Hofn| Iceland|
HFN|BIHN|64.295556|-15.227222|  24| 0| N| Atlantic/Reykjavik|
|  14|             Husavik| Husavik|       Iceland|
HZK|BIHU|65.952328|-17.425978|  48| 0| N| Atlantic/Reykjavik|
|  15|           Isafjordur| Isafjordur|       Iceland|
IFJ|BIIS|66.058056|-23.135278|   8| 0| N| Atlantic/Reykjavik|
|  16|Keflavik Internat...|      Keflavik| Iceland| KEF|BIKF|     63.985|-
22.605556| 171| 0| N| Atlantic/Reykjavik|
|  17|       Patreksfjordur|Patreksfjordur|       Iceland|
PFJ|BIPA|65.555833| -23.965|  11| 0| N| Atlantic/Reykjavik|
|  18|           Reykjavik| Reykjavik|       Iceland| RKV|BIRK| 64.13|-
21.940556|  48| 0| N| Atlantic/Reykjavik|
|  19|         Siglufjordur| Siglufjordur|       Iceland|
SIJ|BISI|66.133333|-18.916667|  10| 0| N| Atlantic/Reykjavik|
+---+--------------------+-------------+----------------+----+----+----
-----+----------+----+--------+----+--------------------+
only showing top 20 rows
```

Now let's add a query. Suppose I want to list all the airports in Ireland. Looking at the previous output, we know that the country is in the _ c3 column. Create a new function called `runShowIrishAirports`.

```java
public static void runShowIrishAirports(SparkSession spark)
throws AnalysisException {
    Dataset<Row> df = spark.read().csv(airportDataPath);
    df.createTempView("airports");
    Dataset<Row> irishAirports = spark
.sql("SELECT _c1, _c4, _c5 FROM airports WHERE _c3='Ireland'");
    irishAirports.show();
}
```

The function reads in the CSV file in the same way as before, but this time we give a temporary view name called `airports` so the raw SQL query has a view to reference. The SQL is a simple `SELECT` statement to read the name of the airport, the respective IATA (the three-letter code), and the ICAO codes. The `.show` method is called `finally`, so the top 20 results are output to the console. Add the new method name to the main method so it is executed at runtime.

Build with Maven as before, which will update the JAR file. You can use the same terminal execution script to run the application. If all is successful, you will see the following output:

```
19/10/22 14:36:00 INFO TaskSchedulerImpl: Removed TaskSet 3.0,
whose tasks have all completed, from pool
19/10/22 14:36:00 INFO DAGScheduler:
ResultStage 3 (show at BasicSparkSQL.java:33) finished in 0.213 s
19/10/22 14:36:00 INFO DAGScheduler: Job 3 finished:
show at BasicSparkSQL.java:33, took 0.222565 s
+--------------------+----+----+
|                 _c1| _c4| _c5|
+--------------------+----+----+
|                Cork| ORK|EICK|
|              Galway| GWY|EICM|
|              Dublin| DUB|EIDW|
|  Ireland West Knock| NOC|EIKN|
|               Kerry| KIR|EIKY|
|            Casement|null|EIME|
|             Shannon| SNN|EINN|
|               Sligo| SXL|EISG|
|           Waterford| WAT|EIWF|
|     Weston Airport|null|EIWT|
|     Donegal Airport| CFN|EIDL|
|   Inishmore Airport| IOR|EIIM|
|Connemara Regiona...| NNR|EICA|
|             Thurles|null|  \N|
|            Limerick|null|  \N|
|            Inisheer| INQ|EIIR|
|              Cashel|null|  \N|
|  Inishmaan Aerodrome| IIA|EIMN|
|               Alpha|null|  \N|
|   Newcastle Airfield|null|EINC|
+--------------------+----+----+
only showing top 20 rows
```

Instead of relying on raw SQL queries, you can do column-based querying by method calls. Consider the following function:

```
public static void runShowIrishAirportsByCols(SparkSession spark)
throws AnalysisException {
   Dataset<Row> df = spark.read().csv(airportDataPath);
   Dataset<Row> filtered = df.filter(col("_c3").contains("Ireland"));
   filtered.show();
}
```

The filter method allows you to create predicate conditions against specific columns. The country column will return the rows that contain the word *Ireland*. The code will show all the columns of the dataset. When executed, it looks like this:

```
19/10/22 14:56:58 INFO TaskSchedulerImpl: Removed TaskSet 1.0,
whose tasks have all completed, from pool
19/10/22 14:56:58 INFO DAGScheduler: ResultStage 1
(show at BasicSparkSQL.java:40) finished in 0.290 s
19/10/22 14:56:58 INFO DAGScheduler: Job 1 finished:
show at BasicSparkSQL.java:40, took 0.293577 s
+----+--------------------+----------+-------+----+----+----------+-----
-----+---+---+----+------------+
| _c0|                  _c1| _c2| _c3| _c4| _c5|        _c6| _c7|_c8|_c9|_
c10| _c11|
+----+--------------------+----------+-------+----+----+----------+-----
-----+---+---+----+------------+
| 596|                Cork| Cork|Ireland| ORK|EICK| 51.841269|
-8.491111|502|  0| E|Europe/Dublin|
| 597|              Galway| Galway|Ireland| GWY|EICM| 53.300175|
-8.941592| 81|  0| E|Europe/Dublin|
| 599|              Dublin| Dublin|Ireland| DUB|EIDW| 53.421333|
-6.270075|242|  0| E|Europe/Dublin|
| 600|  Ireland West Knock| Connaught|Ireland| NOC|EIKN| 53.910297|
-8.818492|665|  0| E|Europe/Dublin|
| 601|               Kerry| Kerry|Ireland| KIR|EIKY| 52.180878|
-9.523783|112|  0| E|Europe/Dublin|
| 602|            Casement| Casement|Ireland|null|EIME| 53.301667|
-6.451333|319|  0| E|Europe/Dublin|
| 603|             Shannon| Shannon|Ireland| SNN|EINN| 52.701978|
-8.924817| 46|  0| E|Europe/Dublin|
| 604|               Sligo| Sligo|Ireland| SXL|EISG| 54.280214|
-8.599206| 11|  0| E|Europe/Dublin|
| 605|           Waterford| Waterford|Ireland| WAT|EIWF|   52.1872|
-7.086964|119| 0| E|Europe/Dublin|
|5578|      Weston Airport|   Leixlip|Ireland|null|EIWT| 53.351333|
-6.4875|150| 0| E|Europe/Dublin|
|5577|     Donegal Airport|   Dongloe|Ireland| CFN|EIDL| 55.044192|
-8.341| 30| 0| E|Europe/Dublin|
|6421|   Inishmore Airport| Inis Mor|Ireland| IOR|EIIM|   53.1067|
-9.65361| 24| 0| U|Europe/Dublin|
|6422|Connemara Regiona...|Indreabhan|Ireland| NNR|EICA|   53.2303|
-9.46778| 0| 0| U|Europe/Dublin|
|6901|             Thurles| Thurles|Ireland|null|  \N| 52.67888|
-7.814369|500| 0|  U|Europe/Dublin|
|6900|            Limerick| Limerick|Ireland|null|  \N| 52.659|
-8.624|500| 0| U|Europe/Dublin|
|7030|            Inisheer| Inisheer|Ireland| INQ|EIIR|   53.0647|
-9.5109| 40| 0| E|Europe/Dublin|
```

```
|7034|                Cashel| Cashel|Ireland|null|  \N|52.5158333|-
7.8855556|440| 0|   U|Europe/Dublin|
|7468| Inishmaan Aerodrome| Inishmaan|Ireland| IIA|EIMN| 53.091944|
-9.57| 13| 0| U|Europe/Dublin|
|8464|                 Alpha| Cork|Ireland|null|  \N| 51.400377|
-7.901464|100| 0|   U| \N|
|8683|  Newcastle Airfield| Newcastle|Ireland|null|EINC| 53.073056|
-6.039722| 14|  0| E|Europe/Dublin|
+----+-------------------+----------+-------+----+----+----------+-----
----+---+---+----+------------+
only showing top 20 rows
```

Wrapping Up SparkSQL

The SparkSQL API is a good way to extract data from large datasets before you start any machine learning. There's little point in using every column of data when you know it's going to provide any value in the final model.

It's perfect for segmenting large volumes of data; it's a big data tool after all. For example, if you were collating data from lots of point-of-sale terminals and you wanted to start running analysis on specific stock item types, using Spark SQL you could pull all the POS data in, query against the type, and then run your machine learning training on the saved output data.

Spark Streaming

In the previous chapter, you learned how Kafka functions as a streaming message log. Within Kafka there are various ways of consuming and transforming the stream of messages. Spark has a streaming API that enables the cluster to process blocks of messages as they are set from a data source. In this section, you will see how the streaming API functions and how it connects with other data sources.

Basic Concepts

Spark Streaming can ingest data from a range of sources, such as ZeroMQ, Kafka, Flume, and raw TCP sockets. As with Spring XD, after data has entered the system, you have the option to process and manipulate the data coming in and then to store it to an outbound location.

Spark Streaming divides data into batches for processing, rather than handling one piece of data at a time like Kafka does. Then Spark Streaming processes and hands those batches to the requested output. Spark calls them *micro batches*.

You can use the raw TCP socket to emit some data and Spark Streaming to ingest it. Spark Streaming uses the concept of a DStream—a discretized stream—which is a continuous stream of data coming in for processing.

Creating Your First Spark Stream

In this example, we will use the Linux nc command to produce data and get Spark Streaming to read it in and do a basic word count. Before code is applied, you will need to add the Spark Streaming library dependency to the Maven pom.xml file. Add the following dependency block to your build file:

```
<dependencies>
.....
 <dependency>
   <groupId>org.apache.spark</groupId>
   <artifactId>spark-streaming_2.11</artifactId>
   <version>2.2.0</version>
 </dependency>
</dependencies>
```

The Java code for the Spark Streaming job is straightforward.

```
package mlbook.ch13.spark;

import org.apache.spark.SparkConf;
import org.apache.spark.api.java.StorageLevels;
import org.apache.spark.streaming.Durations;
import org.apache.spark.streaming.api.java.JavaDStream;
import org.apache.spark.streaming.api.java.JavaPairDStream;
import org.apache.spark.streaming.api.java.JavaReceiverInputDStream;
import org.apache.spark.streaming.api.java.JavaStreamingContext;
import scala.Tuple2;
import java.util.Arrays;
import java.util.regex.Pattern;

public class BasicSparkStreaming {
    private static final Pattern SPACE = Pattern.compile(" ");
    public static void main(String[] args) throws Exception {
        String hostname = "192.168.1.103";
        int port = 9999;

        SparkConf sparkConf = new SparkConf()
.setAppName("BasicSparkStreaming");
        JavaStreamingContext ssc = new JavaStreamingContext(sparkConf,
Durations.seconds(1));

        JavaReceiverInputDStream<String> lines = ssc.socketTextStream(
                hostname, port, StorageLevels.MEMORY_AND_DISK_SER);
```

```
        JavaDStream<String> wordsOnEachLine = lines
    .flatMap(line -> Arrays.asList(SPACE.split(line)).iterator());
        JavaPairDStream<String, Integer> wordCounts = wordsOnEachLine
    .mapToPair(s -> new Tuple2<>(s, 1))
                .reduceByKey((i1, i2) -> i1 + i2);

        wordCounts.print();
        ssc.start();
        ssc.awaitTermination();
    }
}
```

Once again, there are a number of steps for the application to work. `JavaS-treamingContext` sets up the core context with the Spark configuration. The other thing that is set is the duration between the streamed blocks being processed by Spark. In this example, I've set it to one second.

The next step is to set up an input stream with the context, adding the hostname (set it to the IP address of your machine and the port number to whatever you run with `nc`; more on that in a moment). The receiver stream will consume data from the source and then, in the same way as the core Spark word count application, use a flat map on the lines. The split is done on each space character and this is contained with the DStream. The final step is to do the reduce step and add up the keys.

Build the application with Maven as shown previously in this chapter. Once the JAR file is created, you will need to start a fresh terminal window and start `nc`.

If you are using Linux, use the following:

```
nc -lk 9999
```

The `-k` option is not available in macOS, so you will have to omit it.

Once `nc` is running, paste some text into the `nc` window. Nothing will happen apart from seeing the text you pasted in. With the `nc` server running, you can now start the Spark Streaming job.

```
$ /usr/local/spark-2.4.4-bin-hadoop2.7/bin/spark-submit
  --class "mlbook.ch13.spark.BasicSparkStreaming" --master local[4]
target/Chapter13-1.0.0-SNAPSHOT.jar
```

Spark will generate a lot of console output as it will attempt to retrieve data from the source every second as instructed. You will, however, see the word counted stream in the output.

```
19/10/23 09:24:31 INFO Executor: Finished task 0.0 in stage 10.0 (TID
10). 1517 bytes result sent to driver
19/10/23 09:24:31 INFO TaskSetManager: Finished task 0.0 in stage 10.0
(TID 10) in 35 ms on localhost (executor driver) (1/1)
19/10/23 09:24:31 INFO TaskSchedulerImpl: Removed TaskSet 10.0, whose
tasks have all completed, from pool
```

```
19/10/23 09:24:31 INFO DAGScheduler: ResultStage 10 (print at
BasicSparkStreaming.java:34) finished in 0.044 s
19/10/23 09:24:31 INFO DAGScheduler: Job 5 finished: print at
BasicSparkStreaming.java:34, took 0.259986 s
-----------------------------------------
Time: 1571819071000 ms
-----------------------------------------
(interdum,1)
(egestas.,1)
(erat,3)
(faucibus,2)
(sapien,2)
(urna,1)
(pretium,,1)
(Suspendisse,1)
(fringilla.,1)
(laoreet,,1)
...
```

As you paste more text into the nc window, the streaming job will update the
word counts to the console window.

Spark Streams from Kafka

With some minor modifications, it is easy to convert the streaming job to consume
Kafka messages. The setup is straightforward, creating a properties configuration
and adding a consumer group name and a topic (or topics) to consume from.

There is another library that you will have to add to your Maven dependencies.

```
<dependency>
    <groupId>org.apache.spark</groupId>
    <artifactId>spark-streaming-kafka-0-10_2.11</artifactId>
    <version>2.2.0</version>
</dependency>
```

As in the previous chapter, any application consuming from Kafka needs to
know the broker location, a topic name, and a consumer group so the reads can
happen. If a consumer group isn't given, Kafka will assign one, but it's best to
create your own so you are in control; it also makes monitoring easier.

```
String brokers = "localhost:9092";
String topicName = "testtopic";
String groupId = "testtopic-group";
```

Next are the Spark context and configuration; there's no real setup to do here
apart from assigning a name. JavaStreamingContext takes the context configuration and also a duration; for this example, I've set it to five seconds.

```
   SparkConf sparkConf = new SparkConf()
.setAppName("BasicSparkStreamingKafka");
       JavaStreamingContext streamingContext =
new JavaStreamingContext(sparkConf, Durations.seconds(5));
```

The configuration parameters are put into a Java hash map.

```
Map<String, Object> params = new HashMap<>();
       params.put(ConsumerConfig.BOOTSTRAP_SERVERS_CONFIG, brokers);
       params.put(ConsumerConfig.GROUP_ID_CONFIG, groupId);
       params.put(ConsumerConfig.KEY_DESERIALIZER_CLASS_CONFIG,
 StringDeserializer.class);
       params.put(ConsumerConfig.VALUE_DESERIALIZER_CLASS_CONFIG,
 StringDeserializer.class);
```

Topics are specified in a Java Set; there could be more than one topic that is being processed. The string defining the topics should be comma-separated or could just be passed in as a Set.

```
Set<String> topics = new HashSet<>(Arrays
.asList(topicName.split(",")));
```

The input stream is configured as a direct stream to the Kafka cluster. The topic information and Kafka properties are passed in. With this in place, the only things that remain are reading in the Kafka messages consumed, using the DStream to create a stream of lines to be split by spaces, and then mapping/ reducing the words.

```
KafkaUtils.createDirectStream(
               streamingContext,
               LocationStrategies.PreferConsistent(),
               ConsumerStrategies.Subscribe(topics, params));

       JavaDStream<String> lines = messages.map(ConsumerRecord::value);
       JavaDStream<String> words = lines.flatMap(x -> Arrays
.asList(SPACE.split(x)).iterator());
       JavaPairDStream<String, Integer> wordCounts = words
.mapToPair(s -> new Tuple2<>(s, 1))
               .reduceByKey((i1, i2) -> i1 + i2);
       wordCounts.print();
```

MLib: The Machine Learning Library

So far in this book I've covered using machine learning libraries such as Weka, DeepLearning4J, and even Apache Commons Math for some operations. The

Spark framework has a set of machine learning libraries for large-scale learning. It supports a number of algorithm types.

- Logistic regression
- Naïve Bayes
- Decision trees
- Random forests
- K-means clustering
- Association rules
- Latent Dirichlet allocation (LDA) topic modeling

For the remainder of this chapter, I'll show you three machine learning types: decision trees, K-means clustering, and association rules with FP-Growth. Before we can do any of that, we need to add the machine learning library dependencies to the Maven build file.

Dependencies

Open `pom.xml` and add the following dependency with the others we've added for SparkSQL and Spark Streaming APIs:

```
<dependency>
    <groupId>org.apache.spark</groupId>
    <artifactId>spark-mllib_2.11</artifactId>
    <version>2.2.0</version>
    <scope>provided</scope>
</dependency>
```

With the dependencies in place, let's look at the three algorithms that have been covered in previous chapters.

Decision Trees

The basis for Spark jobs is pretty much the same throughout the examples. Create a context, build a configuration, and set a location where the training data is.

The use of the JavaRDD gives us a data placeholder for each step of the learning. With the decision tree, you have to set the number of classes that the algorithm will work with (2 in this example), as well as adding the depth of the tree (with huge volumes of data, you can end up with very large trees).

Passing the RDD data path it will read in the file contents, the next step is to split the data into training and evaluation data. This is done to an RDD array, and this has only two elements, one for the training data and one for the evaluation data.

The work for the entire model is done in one line.

```
DecisionTreeModel model = DecisionTree
.trainClassifier(trainingData, numberOfClasses, categoricalFeaturesInfo,
impurity, maximumTreeDepth, maximumBins);
```

Once complete, the model is evaluated with the evaluation data, and the results are output to the console output. The final code is shown here:

```
package mlbook.ch13.spark;
import java.util.HashMap;
import java.util.Map;
import scala.Tuple2;
import org.apache.spark.SparkConf;
import org.apache.spark.api.java.JavaPairRDD;
import org.apache.spark.api.java.JavaRDD;
import org.apache.spark.api.java.JavaSparkContext;
import org.apache.spark.mllib.regression.LabeledPoint;
import org.apache.spark.mllib.tree.DecisionTree;
import org.apache.spark.mllib.tree.model.DecisionTreeModel;
import org.apache.spark.mllib.util.MLUtils;

public class BasicMLLibDecisionTree {
    public static void main(String[] args) {
        int numberOfClasses = 2;
        Map<Integer, Integer> categoricalFeaturesInfo = new HashMap<>();
        String impurity = "gini";
        int maximumTreeDepth = 7;
        int maximumBins = 48;

        SparkConf sparkConf = new SparkConf()
.setAppName("BasicMLLibDecisionTree");
        JavaSparkContext jsc = new JavaSparkContext(sparkConf);

        String datapath = "/path/to/data/ch13/mllib/dtree.txt";
        JavaRDD<LabeledPoint> data = MLUtils
.loadLibSVMFile(jsc.sc(), datapath).toJavaRDD();
        JavaRDD<LabeledPoint>[] splits = data
.randomSplit(new double[]{0.7, 0.3});
        JavaRDD<LabeledPoint> trainingData = splits[0];
        JavaRDD<LabeledPoint> testData = splits[1];

        DecisionTreeModel model = DecisionTree
.trainClassifier(trainingData, numberOfClasses,
categoricalFeaturesInfo, impurity, maximumTreeDepth, maximumBins);

        JavaPairRDD<Double, Double> predictionAndLabel =
                testData.mapToPair(p -> new Tuple2<>(model
.predict(p.features()), p.label()));
        double predictionTestErrorValue =
```

```
                    predictionAndLabel.filter(pl -> !pl._1().equals(pl._2())))
    .count() / (double) testData.count();

        System.out.println("Prediction test error value: "            +
predictionTestErrorValue);
        System.out.println("Output classification tree:\n" +
model.toDebugString());
    }
}
```

Run Maven to compile and repackage the JAR file with the new application.

```
mvn package
```

Using `spark-submit`, you can then test the application against the training data supplied.

```
$ /usr/local/spark-2.4.4-bin-hadoop2.7/bin/spark-submit \
  --class "mlbook.ch13.spark.BasicMLLibDecisionTree" --master local[4]
target/Chapter13-1.0.0-SNAPSHOT.jar
```

You will see Spark start a local cluster and process the training data. Eventually you will see the output of the decision tree along with the test error value.

```
19/10/24 11:28:54 INFO DAGScheduler: ResultStage 8
(count at BasicMLLibDecisionTree.java:42) finished in 0.015 s
19/10/24 11:28:54 INFO DAGScheduler: Job 6 finished:
count at BasicMLLibDecisionTree.java:42, took 0.017851 s
Prediction test error value: 0.047619047619047616
Output classification tree:
DecisionTreeModel classifier of depth 1 with 3 nodes
  If (feature 406 <= 12.0)
   Predict: 0.0
  Else (feature 406 > 12.0)
   Predict: 1.0
```

Clustering

Spark includes a K-means clustering implementation. Again, the emphasis is on large-scale data. These processing steps during training can be partitioned and clustered over many machines. For this example, I will keep it simple and on the local cluster.

The training data for this example is basic doubles; these will be read and clustered into three cluster classifications. The data has been kept isolated to illustrate the clusters clearly.

```
0.0 0.0 0.0
0.1 0.1 0.1
0.2 0.2 0.2
```

```
5.0 5.1 5.1
5.2 5.0 5.2
5.0 5.3 5.4
5.0 5.1 5.3
9.0 9.0 9.0
9.1 9.1 9.1
9.2 9.2 9.2
```

The code is made up of the following steps. First, the test data is loaded and split by the space character. This is then iterated and stored in a `Vector` class. Each `Vector` RSS contains a dense vector of each of the values from that line.

Once the number of desired clusters and number of iterations to train the model are set, it's a case of creating the model. This is done in one line, passing in the RDD of vectors, the number of clusters required, and the iteration count.

```
KMeansModel kMeansClusters = KMeans.train(parsedData.rdd(),
  numberOfClassClusters, numberOfIterations);
```

The final steps of the application are to report the findings of the model training. The cost value is calculated and output to the console; then the full clusters are displayed.

```
package mlbook.ch13.spark;
import org.apache.spark.SparkConf;
import org.apache.spark.api.java.JavaSparkContext;
import org.apache.spark.api.java.JavaRDD;
import org.apache.spark.mllib.clustering.KMeans;
import org.apache.spark.mllib.clustering.KMeansModel;
import org.apache.spark.mllib.linalg.Vector;
import org.apache.spark.mllib.linalg.Vectors;

public class BasicMLLibKMeans {
      public static void main(String[] args) {
            SparkConf sparkConf = new SparkConf()
.setAppName("BasicMLLibKMeans");
            JavaSparkContext sparkContext = new
JavaSparkContext(sparkConf);
            JavaRDD<String> data = sparkContext
.textFile("/path/to/data/ch13/mllib/kmeans.txt");
            JavaRDD<Vector> parsedData = data.map(s -> {
                String[] sarray = s.split(" ");
                double[] values = new double[sarray.length];
                for (int i = 0; i < sarray.length; i++) {
                    values[i] = Double.parseDouble(sarray[i]);
                }
                return Vectors.dense(values);
            });
            parsedData.cache();
```

```
          int numberOfClassClusters = 3;
          int numberOfIterations = 50;
          KMeansModel kMeansClusters = KMeans.train(parsedData.rdd(),
      numberOfClassClusters, numberOfIterations);

          double cost = kMeansClusters.computeCost(parsedData.rdd());
          System.out.println("Computed cost: " + cost);

          System.out.println("Showing cluster centres: ");
          for (Vector center: kMeansClusters.clusterCenters()) {
              System.out.println(" " + center);
          }

          sparkContext.stop();
      }
  }
```

Again, run Maven to compile and repackage the JAR file with the new K-means application code.

```
mvn package
```

Using `spark-submit`, you can then test the application against the training data supplied.

```
$ /usr/local/spark-2.4.4-bin-hadoop2.7/bin/spark-submit \
--class "mlbook.ch13.spark.BasicMLLibKMeans" --master local[4] target/
Chapter13-1.0.0-SNAPSHOT.jar
```

Spark will go through the usual startup process and then will output the cost function and the clusters.

```
19/10/24 11:48:56 INFO DAGScheduler: ResultStage 11
(sum at KMeansModel.scala:105) finished in 0.018 s
19/10/24 11:48:56 INFO DAGScheduler: Job 9 finished:
sum at KMeansModel.scala:105, took 0.021366 s
19/10/24 11:48:56 INFO TorrentBroadcast: Destroying Broadcast(17)
(from destroy at KMeansModel.scala:106)
Computed cost: 0.24749999999988637
Showing cluster centres:
  [5.05,5.125,5.25]
  [0.1,0.1,0.1]
  [9.1,9.1,9.1]
```

Association Rules with FP-Growth

As we've previously seen, it's possible to suggest items for a customer based on historical shopping basket contents. Spark has an implantation of the FP-Growth algorithm within Spark MLLib to do basket (or any other data) analysis at volume.

First, I need some shopping basket transactions. This is just a simple text file of items; each line represents the contents of one basket checked out.

```
milk tea fruit flour pencils
tea cake biscuits beans peas hot_chocolate eggs coffee
coffee biscuits pasta newspaper milk
biscuits tea cake soap eggs coffee paper books
tea
biscuits tea cake milk paper eggs pencils
```

The supporting Spark code does similar functions as to the decision tree and K-means examples. The context is set up, and then the data is read and parsed into an RDD.

For FP-Growth, we need to set the minimum support and the number of partitions to create while the associations are computed.

As you may remember from the previous association rule example in Weka, we want to show only the associations with a minimum confidence level. For this example, it's set to 70 percent (set as a double value of 0.7 in the code).

The final step is to generate the association rules and output them to the console. We also want to see the antecedent item and the item consequent along with the confidence score for that rule.

```
package mlbook.ch13.spark;
import java.util.Arrays;
import java.util.List;
import org.apache.spark.api.java.JavaRDD;
import org.apache.spark.api.java.JavaSparkContext;
import org.apache.spark.mllib.fpm.AssociationRules;
import org.apache.spark.mllib.fpm.FPGrowth;
import org.apache.spark.mllib.fpm.FPGrowthModel;
import org.apache.spark.SparkConf;

public class BasicMLLibFPGrowth {
    public static void main(String[] args) {
        SparkConf conf = new SparkConf().setAppName("BasicMLLibFPGro
wth");
        JavaSparkContext sc = new JavaSparkContext(conf);

        JavaRDD<String> data = sc
.textFile("/path/to/data/ch13/mllib/fpgrowth_items.txt");
        JavaRDD<List<String>> basketItems = data.map(line -> Arrays
.asList(line.split(" ")));

        FPGrowth fpg = new FPGrowth()
                .setMinSupport(0.2)
                .setNumPartitions(10);
        FPGrowthModel<String> model = fpg.run(basketItems);
```

```
        for (FPGrowth.FreqItemset<String> itemset: model.freqItemsets()
.toJavaRDD().collect()) {
            System.out.println("(" + itemset.javaItems() + "), " +
itemset.freq());
        }

        double minConfidence = 0.7;
        for (AssociationRules.Rule<String> rule
                : model.generateAssociationRules(minConfidence)
.toJavaRDD().collect()) {
            System.out.println(
                    rule.javaAntecedent() + " => " +
rule.javaConsequent() + ", " + rule.confidence());
        }
        sc.stop();
    }
}
```

For one last time, run Maven to compile and repackage the JAR file with the association rules application code.

```
mvn package
```

Using `spark-submit`, you can then test the application against the shopping basket data.

```
$ /usr/local/spark-2.4.4-bin-hadoop2.7/bin/spark-submit \
  --class "mlbook.ch13.spark.BasicMLLibFPGrowth" --master local[4]
target/Chapter13-1.0.0-SNAPSHOT.jar
```

Spark will start, read in the data, and compute the association rules model. Once complete, the results of the model will be output to the console.

```
19/10/24 11:01:03 INFO TaskSchedulerImpl: Removed TaskSet 8.0,
 whose tasks have all completed, from pool
19/10/24 11:01:03 INFO DAGScheduler: ResultStage 8
(collect at BasicMLLibFPGrowth.java:32) finished in 0.144 s
19/10/24 11:01:03 INFO DAGScheduler: Job 3 finished:
collect at BasicMLLibFPGrowth.java:32, took 0.485325 s
[cake, eggs, coffee] => [biscuits], 1.0
[cake, eggs, coffee] => [tea], 1.0
[paper, eggs, tea] => [cake], 1.0
[paper, eggs, tea] => [biscuits], 1.0
[eggs, coffee, tea] => [biscuits], 1.0
[eggs, coffee, tea] => [cake], 1.0
[coffee] => [biscuits], 1.0
[paper, cake, biscuits] => [tea], 1.0
[paper, cake, biscuits] => [eggs], 1.0
[eggs, coffee] => [biscuits], 1.0
[eggs, coffee] => [tea], 1.0
```

```
[eggs, coffee] => [cake], 1.0
[eggs, biscuits] => [tea], 1.0
[eggs, biscuits] => [cake], 1.0
```

Summary

The Spark framework has changed a lot since it was covered in the first edition of this book. The APIs are a lot easier to code with, especially for Java. If you use Python, then it's worth looking at the PySpark REPL for doing Spark work. If you are a Clojure user, then there are a number of Spark wrappers available to you as well.

Remember, this framework is for large-scale data. If you have medium or small amounts of data, then look at the alternatives that were shown earlier in the book first. If those seem to struggle while training, then it's worth looking at Spark as the next stage of your development pipeline.

Machine Learning with R

When you're in a room of data scientists, statisticians, and math types, you'll hear one letter crop up again and again: the letter *R*. R is a programming language, and it's basically command-line driven. In addition to being used in the command-line shell, R can be written in code form and run.

Why am I telling you all this? Well, on top of the programming skills that get mentioned, you might also be asked, "Do you do R?" After this chapter, you'll hopefully have a starting point to reply, "Yes!"

Installing R

The R language comes ready to use for a number of operating systems. The download page at http://www.r-project.org has a number of mirror sites, so pick a mirror that's closest to you. From the mirror, choose the download for your operating system.

macOS

The current version of R (3.6 at time of writing) will run on the 64-bit Intel-based Macs. Download the file and open it to install it. It installs the R binaries into the /Applications folder.

Windows

The .exe download for Windows provides binaries for running on 32- or 64-bit machines. The base package download will provide you with everything you need to get started.

Linux

Binary downloads are available for Debian, Ubuntu, Red Hat, and SUSE Linux distributions. If you want to save some time (and effort) and you're running Debian or Ubuntu, then you can use apt-get to install the r-base and r-base-dev packages. Ensure that the repository package base is up-to-date first. For users of the Red Hat family of distributions, use the command sudo yum install R.

Your First Run

When you run R, you're presented with the basic R shell, as shown in Figure 14.1. This is the main place where the work is done. It's sparse, but it does the job fine.

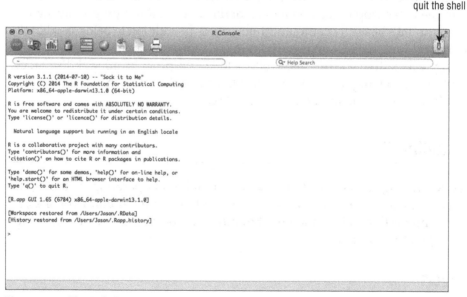

Figure 14.1: The R shell

If at any time you want help on a topic, you can use the `help` command. For example, if you want to know about Standard Deviation, just type **help(sd)**, and R opens a new window with the information (see Figure 14.2).

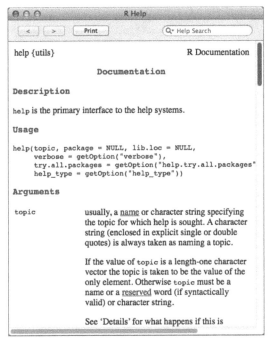

Figure 14.2: R's help system

You can quit the shell either by clicking the light switch on the top right of the program window (refer to Figure 14.1) or by typing **quit()** on the command line.

For basic needs, the R shell is fine and does the job well. For an actual development environment, you have to install some more software such as R-Studio.

Installing R-Studio

The R-Studio project (see Figure 14.3) is a commercial integrated development environment (IDE) for R. It comes in an open source community edition that is free to use. To download R-Studio IDE, visit the following website and select your operating system type:

```
http://www.rstudio.com/products/rstudio/download
```

Make sure that the R base binary is installed as described in the preceding section before you download R-Studio.

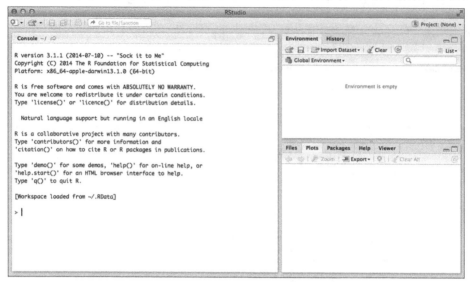

Figure 14.3: R-Studio

The R Basics

To run through the R basics, I'm going to use the standard R development environment. The command-line prompt is a simple greater-than sign (>).

You can perform calculations on the command line, so adding numbers together is a trivial process, like so:

```
> 1+2
[1] 3
>
```

To get proper use from R, though, you need to think a little more programmatically.

Variables and Vectors

R supports variables as you would expect. To assign them, you can either use the equal sign (=) or the less-than sign and a hyphen together (<-):

```
> myage = 21
> myageagain <- 21
```

```
> myage
[1] 21
> myageagain
[1] 21
>
```

Variables can also store string variables and other data types. The ones you'll use most are numeric values.

Lists of data are held in arrays, called *vectors* in R, and are defined with the `c()` function.

```
> lotterynums <- c(2,7,20,35,36,42)
> lotterynums
[1]  2  7 20 35 36 42
```

Vectors can also hold strings. Using the `length()` function tells you how many elements are in the array.

```
> kc <- c("Robert", "Jakko", "Tony", "Mel", "Pat",
 "Jeremy","Gavin","Bill")
> kc
[1] "Robert" "Jakko" "Tony" "Mel" "Pat" "Jeremy" "Gavin" "Bill"
> length(kc)
[1] 8
>
```

To show specific values in the array, you can use the variable name and the element you want to show.

```
> kc[5]
[1] "Pat"
```

Matrices

Now that you know how vector lists of numbers work, you can convert them into a matrix. To define a matrix, you take the data and then define how many rows and columns you require.

```
> mymatrix <- matrix(c(1,2,3,4,5,6,7,8,9,10), nrow=2, ncol=5,
  byrow=TRUE)
> mymatrix
     [,1] [,2] [,3] [,4] [,5]
[1,]    1    2    3    4    5
[2,]    6    7    8    9   10
>
```

You can then retrieve data based on the row and column position.

```
> # by row, col
> mymatrix[2,4]
[1] 9
> # entire row
> mymatrix[2,]
[1]  6  7  8  9 10
> # entire col
> mymatrix[,4]
[1] 4 9
>
```

Instead of numeric row and column names, you can define text label names to make things more readable.

```
> dimnames(mymatrix) <- list(c("row1","row2"),c("c1","c2",
  "c3","c4","c5"))
> mymatrix
     c1 c2 c3 c4 c5
row1  1  2  3  4  5
row2  6  7  8  9 10
>
```

You can reference data by row and column by using the row and column names you've just defined.

```
> mymatrix["row2", "c5"]
[1] 10
```

Lists

A *list* is a vector containing other objects. This can be a mixture of objects (numeric, Boolean, and strings, for example) or other vectors within the list.

```
> nums <- c(1,2,3,4,5)
> strings <- c("hello", "world", "again")
> bools <- (TRUE, FALSE)
Error: unexpected ',' in "bools <- (TRUE,"
> bools <- c(TRUE, FALSE)
> mylist <- list(bools, strings, nums)
> mylist
[[1]]
[1]  TRUE FALSE

[[2]]
[1] "hello" "world" "again"

[[3]]
[1] 1 2 3 4 5
```

To retrieve the strings on their own, you can slice the list accordingly with the `[]` notation.

```
> mylist[2]
[[1]]
[1] "hello" "world" "again"
```

To reference a member of the listed object directly, you have to use a double squared bracket. You can modify the member within the list as well.

```
> mylist[[2]][1]
[1] "hello"
> mylist[[2]][1] <- "goodbye"
> mylist
[[1]]
[1]   TRUE FALSE

[[2]]
[1] "goodbye" "world"    "again"

[[3]]
[1] 1 2 3 4 5

>
```

Data Frames

Data frames are basically lists of vectors. The column count is the same in the vectors. R comes with some predefined data frames to play with. Using the `head()` function, you can see the top few lines of the data frame. This saves the entire contents of the frame being shown in the command line.

```
> data(USArrests)
> head(USArrests)
           Murder Assault UrbanPop Rape
Alabama      13.2     236       58 21.2
Alaska       10.0     263       48 44.5
Arizona       8.1     294       80 31.0
Arkansas      8.8     190       50 19.5
California    9.0     276       91 40.6
Colorado      7.9     204       78 38.7
```

You can reference data with the row and column positioning like you did with the matrices.

```
> USArrests["New York",]
         Murder Assault UrbanPop Rape
New York   11.1     254       86 26.1
> USArrests["New York", "Assault"]
[1] 254
```

Installing Packages

R comes with a comprehensive selection of packages that are available to download. You can see the Comprehensive R Archive Network (usually referred to as CRAN) packages that are available on the R website at www.r-project.org/. They are a broad spectrum of statistics, data-processing, and other tools.

To install the packages, you use the `install.packages()` function from the R command line. It takes care of everything for you.

For example, to install the tools for Approximate Bayesian Computation (ABC), you install the `abc` package:

```
>install.packages("abc")
```

Some packages might require dependencies to be installed first, so it's prudent to use the `dependencies` flag to ensure they are installed, too.

```
>
also installing the dependencies 'SparseM', 'quantreg', 'locfit'

trying URL 'http://cran.rstudio.com/bin/macosx/mavericks/contrib/3.1/
SparseM_1.03.tgz'
Content type 'application/x-gzip' length 825491 bytes (806 Kb)
opened URL
==================================================
downloaded 806 Kb

trying URL 'http://cran.rstudio.com/bin/macosx/mavericks/contrib/3.1/
quantreg_5.05.tgz'
Content type 'application/x-gzip' length 1846783 bytes (1.8 Mb)
opened URL
==================================================
downloaded 1.8 Mb

trying URL 'http://cran.rstudio.com/bin/macosx/mavericks/contrib/3.1/
locfit_1.5-9.1.tgz'
Content type 'application/x-gzip' length 597404 bytes (583 Kb)
opened URL
==================================================
downloaded 583 Kb

trying URL 'http://cran.rstudio.com/bin/macosx/mavericks/contrib/3.1/
abc_2.0.tgz'
Content type 'application/x-gzip' length 5303210 bytes (5.1 Mb)
opened URL
==================================================
downloaded 5.1 Mb
```

```
The downloaded binary packages are in
/var/folders/b5/fz_57qk522nd6vqk2pd4lytr0000gn/T//RtmpOqHaEV/
downloaded_packages
```

To use the library after it's installed, you call it with the `library()` function. It initializes and gives notice of the dependencies it has also loaded.

```
> library(abc)
Loading required package: nnet
Loading required package: quantreg
Loading required package: SparseM

Attaching package: 'SparseM'

The following object is masked from 'package:base':

    backsolve

Loading required package: MASS
Loading required package: locfit
locfit 1.5-9.1 2013-03-22
```

Loading in Data

With the basic notions of variables, lists, and vectors in place, it's time to look at getting some data loaded into R.

CSV Files

The `read.csv` function reads a `.csv` file and loads it into a data frame.

```
> trans <- read.csv('vdata.csv', header=TRUE, sep=',')
> head(trans)
  wheels chassis pax vtype
1      4       2   4   Car
2      9      20  25   Bus
3      5      14  18   Bus
4      5       2   1   Car
5      9      17  25   Bus
6      1       1   1  Bike
>
```

If your .csv file has the column names in the first line, then use header=TRUE; otherwise, set it to FALSE. The separator is defined with the sep keyword, and you define whatever delimiter you want. If you have missing values, then it's wise to use the fill flag as well to ensure that your data will have the correct number of elements in each row.

MySQL Queries

Installing the RMySQL package gives you access to MySQL databases. You can pull queries into R so they can be processed.

```
>install.packages("RMySQL", dependencies=TRUE)
```

If you are working on a Windows-based system, then the library requires building for the source files. Two environment variables are required for the library to compile.

```
> Sys.setenv(PKG_CPPFLAGS = "-I/path/to/mysql/include/dir")
> Sys.setenv(PKG_LIBS = "-L/path/to/library/dir -lmysqlclient")
> install.packages("RMySQL", type = "source")
```

As with Java code, you need to define a connection to the database before you can query it.

```
> con <- dbConnect(MySQL(), user="myuser", password="mypass",
  dbname="mydb", host="localhost")
```

From there, after you have a connection, you can see what tables are in the database.

```
> dbListTables(con)
```

You then query a table. The data from the query is returned as a data frame.

```
>dta <- dbGetQuery(con, "SELECT * FROM mytable")
```

There are a number of other databases supported in R, including SQLite3, PostgreSQL, and Oracle.

Creating Random Sample Data

Perhaps you don't have any data to load or you just want to have a random sample of numbers to play with. Using the sample function, you can create a handy vector of numbers.

```
> sam <- sample.int(1000, 20, replace=TRUE)
> sam
 [1]   32 192 783 654 250 261 150 687 619 332 549 225 545 175
508 782 237 748 334 804
```

Plotting Data

R supports basic plots of your data. They can take a little amount of getting use to with regard to the syntax, so the following sections provide a short primer.

Bar Charts

How many bar charts did you draw at school? I drew far more than I care to remember, but R makes it easy for me now. (See Figure 14.4.)

```
> sam <- sample.int(1000, 20, replace=TRUE)
> sam
 [1]   32 192 783 654 250 261 150 687 619 332 549 225 545 175
508 782 237 748 334 804
> barplot(sam, main="My first plot", horiz=TRUE)
```

Figure 14.4: Horizontal bar chart

If you remove the `horiz` option, you get the bars traveling in a vertical direction, as shown in Figure 14.5.

```
> barplot(sam, main="My first plot")
```

Pie Charts

The pie charts in R are basic (see Figure 14.6), but they get the job done. It's just a case of giving the pie chart values and labels. You can easily expand on this if necessary.

```
> pie(sam, main="First Pie Chart", labels=sam)
```

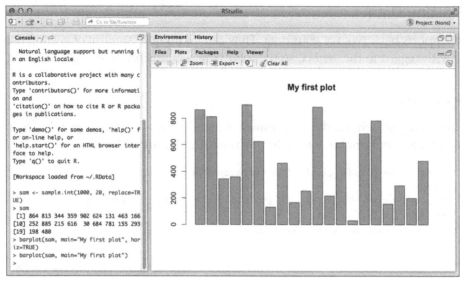

Figure 14.5: Vertical bar chart

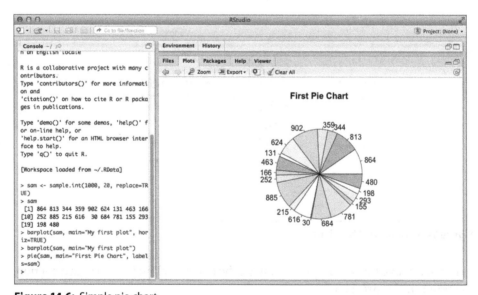

Figure 14.6: Simple pie chart

Dot Plots

The dot plot function (see Figure 14.7) is a simple case of specifying a vector. You can also group the dot plot into specific sections if required.

```
> dotchart(sam, main="My Dot Chart", labels="Value", xlab="Frequency")
```

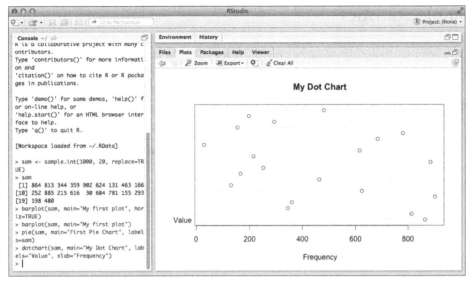

Figure 14.7: Simple dot plot

Line Charts

With two vectors of numbers, you can create a line chart, as shown in Figure 14.8.

```
> sam1 <- sample.int(10, 12, replace=TRUE)
> sam1
 [1]  5 10  4  8  2  2  2  4  2  5  2  6
> sam2 <- sam1
> plot(sam1, sam2)
> lines(sam1, sam2, type="l")
```

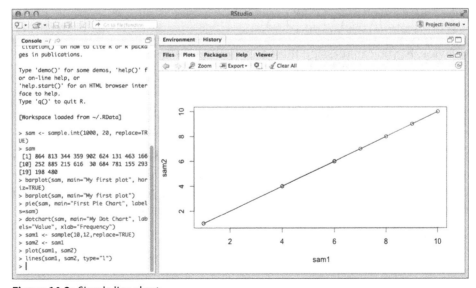

Figure 14.8: Simple line chart

Simple Statistics

R is about statistics; that's what it's built for. Unlike Java, Scala, or Python, R's syntax is a little unforgiving, but after a few sessions, it becomes more natural.

Try creating a simple vector of numbers, and then you can work through some functions. Start with the basics.

```
> s <- sample(100, 12, replace=TRUE)
> # get a basic summary of the vector: lowest value, 1st quartile,
 median, mean, 3rd quartile and maximum value
> summary(s)
   Min. 1st Qu.  Median    Mean 3rd Qu.    Max.
   1.00    9.25   28.50   37.58   58.25   97.00
> # just get the minimum
> min(s)
[1] 1
> # get the maximum value
> max(s)
[1] 97
> # get the average
> mean(s)
[1] 37.58333
> # get the median
> median(s)
> # get the standard deviation
> sd(s)
[1] 31.57807
> # use the table function to see the frequency of the data
> table(s)
s
 1  5  7 10 22 25 32 55 57 62 78 97
 1  1  1  1  1  1  1  1  1  1  1  1
```

Obviously, you can reassign these function results as new variables or vectors. This gives you the basic outline of how the summaries work.

Simple Linear Regression

This section gives an example of simple linear regression in R. It will give you a good idea of how things are put together. Here's the story: you have made profit based on the number of seconds that the sales team is on a call. If you know the profit made, can you calculate how long the call took?

Creating the Data

First, create two separate vectors: one for the number of seconds in the call (`secondsCall`) and another for the amount of profit that was made (`dollarProfit`).

```
> # setup the data
> secondsCall <- c(23,28,39,48,64,75,88,96,97,109,118,149,150,156,165)
> dollarProfit <- c(1,2,3,3,4,4,5,6,6,7,8,8,9,10,10)
```

The Initial Graph

You can create a simple plot for those values (see Figure 14.9) by using the `plot` command.

```
> # create a simple plot
> plot(secondsCall, dollarProfit)
```

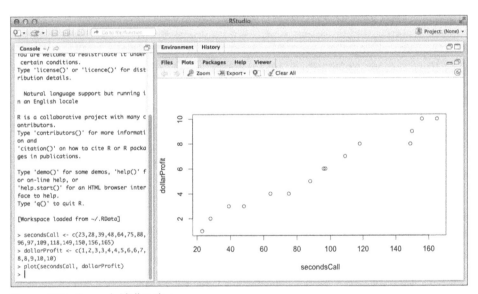

Figure 14.9: Seconds/dollar plot

Regression with the Linear Model

Within R, there is a command that will do the linear model for you: `lm`. You can define the model and save it as a variable. The order of variables is dependent (`secondsCall`), followed by a tilde symbol (~), and finally the independent variables (`dollarProfit`).

```
> # define the linear model
> model <- lm(secondsCall ~ dollarProfit)
```

```
> model

Call:
lm(formula = secondsCall ~ dollarProfit)

Coefficients:
  (Intercept)   dollarProfit
       0.6226        16.2286

>
```

Now that you know the intercept (0.6226) and the dollar profit amount of 16.22, you can expand on the model information using the summary command.

```
> # expend the summary of the model
> summary(model)

Call:
lm(formula = secondsCall ~ dollarProfit)

Residuals:
     Min      1Q  Median      3Q     Max
 -12.451  -5.151  -1.308   4.734  18.549

Coefficients:
              Estimate Std. Error t value Pr(>|t|)
(Intercept)     0.6226     4.8981   0.127    0.901
dollarProfit   16.2286     0.7681  21.129  1.9e-11 ***
---
Signif. codes:  0 '***' 0.001 '**' 0.01 '*' 0.05 '.' 0.1 ' ' 1

Residual standard error: 8.306 on 13 degrees of freedom
Multiple R-squared:  0.9717, Adjusted R-squared:  0.9695
F-statistic: 446.4 on 1 and 13 DF,  p-value: 1.898e-11
```

So, you have a basic model that gives you a regression equation of *secondsCall* = 0.6226 + 16.2268 * *profit amount*. To put the regression line on the plot, use the abline function.

```
> abline(model)
```

Making a Prediction

Assume that someone made a $5 profit, and you want to know the duration of the call based on the model you've just created. Using the predict command, you can make the prediction.

```
> # make a basic prediction of someone making $5.
> predict(model, newdata=data.frame(dollarProfit=5))
```

```
      1
81.76568
>
```

The prediction is that the person was on a call for 81 seconds. You can extend that by adding different `interval` types, which will give you the upper and lower prediction amounts based on the model.

```
> predict(model, newdata=data.frame(dollarProfit=5), interval="pred")
       fit      lwr       upr
1 81.76568 63.19372 100.3376
> predict(model, newdata=data.frame(dollarProfit=5),
 interval="confidence")
       fit      lwr       upr
1 81.76568 76.97553 86.55583
```

Basic Sentiment Analysis

Mining Twitter data and measuring the positive or negative sentiment can be done in R easily. You will require an application to be registered on your Twitter Developer account. The text to rate could be anything from simple sentences typed in to reading in a Twitter stream or a file.

To learn how to create a Twitter Developer application with the required tokens and keys, please refer to Appendix B. There is a full walk-through on how to do this. Once you have the required tokens and keys, you can continue.

Using Functions to Load in Word Lists

You need two sets of text files: one with the positive words and one with the negative words. You can write two quick functions to load the text files and save them to two separate lists.

```
LoadPosWordSet<-function(){
 iu.pos = scan("positive-words.txt", what='character', comment.char=";")
 pos.words = c(iu.pos)
 return(pos.words)
}
```

Then you do the same for the negative word list:

```
LoadNegWordSet<-function(){
 iu.neg = scan("negative-words.txt", what='character', comment.char=";")
 neg.words = c(iu.neg)
 return(neg.words)
}
```

Writing a Function to Score Sentiment

You have a function that takes in a sentence and two word lists (positive and negative sentiment words). Now you can test it.

```
GetScore<-function(sentence, pos.words, neg.words) {
  sentence = gsub('[[:punct:]]', '', sentence)
  sentence = gsub('[[:cntrl:]]', '', sentence)
  sentence = gsub('\\d+', '', sentence)

  sentence = tolower(sentence)

  word.list = str_split(sentence, '\\s+')
  words = unlist(word.list)

  pos.matches = match(words, pos.words)
  neg.matches = match(words, neg.words)

  pos.matches = !is.na(pos.matches)
  neg.matches = !is.na(neg.matches)
  score = sum(pos.matches) - sum(neg.matches)

  return(score)
}
```

The first thing that happens is the sentence is cleaned up with punctuation, control characters, and numbers removed. That should give you just a sentence of words; you then convert it into all lowercase letters.

You split the sentence into a list of words and find out how many times the words match in the positive word list. You also do the same with the negative word list.

With a positive score and negative score, you take the negative away from the positive to get the final score.

You can save the functions as an R source file. You can either create it in a text editor or use the R-Studio editor to create the source file. For this example, I save it all in a file called sentiment.r.

Testing the Function

To test the sentiment code, you first need to load the code and the required library into R.

```
>install.packages("stringr")
>library(stringr)
>source('sentiment.r')
```

```
> pos.words <- LoadPosWordSet()
Read 2006 items

> neg.words <- LoadNegWordSet()
Read 4783 items
```

Now you have 2,006 positive words and 4,783 negative words loaded. By using the GetScore method, you can get a score now on some text, and you can make up some to test. For example, here's a positive one:

```
> testscore<-GetScore("This concert is the best thing I've been to!",
 pos.words, neg.words)
> testscore
[1] 1
```

As you can see, the sentiment analysis gave a score of +1, so it's positive. Try a negative sentence:

```
> testscore2<-GetScore("That's bad real bad, horrible", pos.words,
 neg.words)
> testscore2
[1] -3
```

With a negative string, you get a score of −3. With this basic function, you could process a list of sentences and create a bar graph of the scoring.

Apriori Association Rules

With a set of transactions, you can run a basic Apriori algorithm. The R base system requires a package called arules to be installed before use.

Installing the arules Package

Before you get started, you have to install the arules package.

```
> install.packages("arules", dependencies=TRUE)
also installing the dependencies 'colorspace', 'TSP', 'gclus',
'scatterplot3d', 'vcd', 'seriation', 'igraph', 'pmml', 'XML',
'arulesViz', 'testthat'
The downloaded binary packages are in
/var/folders/b5/fz_57qk522nd6vqk2pd4lytr0000gn/T//Rtmpgp3zNQ/downloaded_
packages
> library(arules)
Loading required package: Matrix
```

```
Attaching package: 'arules'

The following objects are masked from 'package:base':
%in%, write

>
```

Gathering the Training Data

I have prepared a basic .csv file with the basket ID and one item per line. You can see that there are repeating basket IDs to show that the basket contains multiple items. I've called my file transactions.csv.

```
1001,Fries
1001,Coffee
1001,Milk
1002,Coffee
1002,Fries
1003,Coffee
1003,Coke
1003,Eraser
1004,Coffee
1004,Fries
1004,Cookies
1005,Milk
1006,Coffee
1006,Milk
1007,Coffee
1007,Fries
1008,Fries
1008,Coke
```

Importing the Transaction Data

With the read.transactions method, you load the .csv data into a transactions object. It works in a similar way to the read.csv function you saw at the start of the chapter.

I'm setting the rm.duplicates flag to FALSE, because I don't want basket items to be removed.

```
> transactions <- read.transactions(file="transactions.csv",
rm.duplicates=FALSE, format="single", sep=",", cols=c(1,2))
> transactions
transactions in sparse format with
 8 transactions (rows) and
 6 items (columns)
>
```

As you can see, the `transaction` object knows there are eight transactions with six items. You can see the relative frequency of items graphically by using the `itemFrequencyPlot` function, which generates a graph like the one in Figure 14.10.

```
> itemFrequencyPlot(transactions)
```

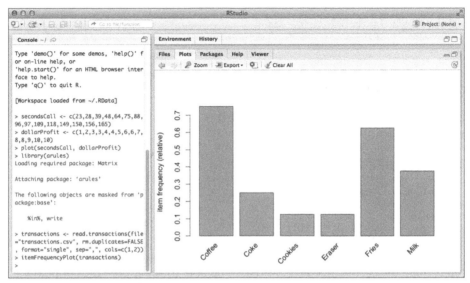

Figure 14.10: Transaction frequencies

Running the Apriori Algorithm

The function to run the algorithm is done in one line. You supply the transaction objects and a set of parameters. When you run the algorithm, you're presented with the resulting output. So, with a support of 0.5 and a confidence of 0.8 (the system is 80 percent confident), you get one association rule.

```
> minedbasketrules <- apriori(transactions, parameter=list(sup=0.5,
conf=0.8, target="rules"))

parameter specification:
 confidence minval smax arem  aval originalSupport support minlen maxlen
target    ext
      0.8    0.1    1 none FALSE            TRUE    0.5      1     10
rules FALSE

algorithmic control:
 filter tree heap memopt load sort verbose
    0.1 TRUE TRUE  FALSE TRUE    2    TRUE
```

```
apriori - find association rules with the apriori algorithm
version 4.21 (2004.05.09)        (c) 1996-2004    Christian Borgelt
set item appearances ...[0 item(s)] done [0.00s].
set transactions ...[6 item(s), 8 transaction(s)] done [0.00s].
sorting and recoding items ... [2 item(s)] done [0.00s].
creating transaction tree ... done [0.00s].
checking subsets of size 1 2 done [0.00s].
writing ... [1 rule(s)] done [0.00s].
creating S4 object  ... done [0.00s].
>
```

Inspecting the Results

Take a look at the one rule and see what it is. You use the `inspect` command to look at the result.

```
> inspect(minedbasketrules)
  lhs          rhs        support confidence     lift
1 {Fries} => {Coffee}      0.5          0.8 1.066667
>
```

There's an 80 percent chance that if someone buys fries, they'll also buy a coffee. If you had thousands or tens of thousands of transactions, then you would raise the confidence level and also be able to see more rules appearing.

Accessing R from Java

Like the power of the Weka workbench can be accessed from a Java program, so too can R code. With the `rJava` bridge, you can run R within Java code and Java within R code.

Installing the rJava Package

Originally the `rJava` package was split along with the JRI package; installing them was a technical process at the time. Fortunately, that has all been replaced with a single binary that covers both packages.

```
> install.packages("rJava")
trying URL 'http://cran.rstudio.com/bin/macosx/mavericks/contrib/3.1/
rJava_0.9-6.tgz'
Content type 'application/x-gzip' length 600621 bytes (586 Kb)
opened URL
==================================================
downloaded 586 Kb
```

```
The downloaded binary packages are in
/var/folders/b5/fz_57qk522nd6vqk2pd4lytr0000gn/T//Rtmpgp3zNQ/downloaded_
packages
>
```

The rJava package uses JNI to talk to Java libraries. From the point of view of working within R, things might seem a little cumbersome, but they do work fine.

Creating Your First Java Code in R

Open your R console or R-Studio. Assuming you've installed the package as described earlier in this chapter, you can do the following:

```
> library(rJava)
> .jinit()
> stringobj <- .jnew("java/lang/String", "This is a string as a Java
object, in R!")
> stringobj
[1] "Java-Object{This is a string as a Java object, in R!}"
```

After the library is loaded, you need to initialize the rJava system with the .jinit() method. You then create a new variable in R that is going to contain a Java string object. The .jnew() method creates a new String object and populates the string. Notice that you have to put the full Java package name in with a slashed notation and not the dotted one.

If you want to find the location of the word *Java* in the string, you use Java's indexOf method. You can call it with rJava by executing the following:

```
[1] "Java-Object{This is a string as a Java object, in R!}"
> .jcall(stringobj, "I", "indexOf", "Java")
[1] 22
```

The command looks involved. The first parameter is the existing object you previously created. The next is the return type from method; because the indexOf method returns an integer, you use the "I" in the calling method. Next is the method name—"indexOf"—and last is the thing you're looking for, "Java". You see the result on the line underneath.

For the full package information for the rJava interface, take a look at the method list at

```
http://rforge.net/doc/packages/rJava/00Index.html
```

Calling R from Java Programs

The interface for calling R from Java is called JRI. The files required to do this are all in the library that was installed from R. There are two components that your Java project requires: the jar file (called JRI.jar) and the native library file (the name changes depending on the operating system you are using—on macOS, it's called `libjri.jnilib`).

```
Jason-Bells-MacBook-Pro:jri Jason$ pwd
/Library/Frameworks/R.framework/Resources/library/rJava/jri
Jason-Bells-MacBook-Pro:jri Jason$ ls -l
total 256
-rw-r--r--  1 Jason   admin   31384 24 Apr 16:02 JRI.jar
-rw-r--r--  1 Jason   admin   10272 24 Apr 16:02 JRIEngine.jar
-rw-r--r--  1 Jason   admin   32354 24 Apr 16:02 REngine.jar
drwxr-xr-x  8 Jason   admin     272 24 Apr 16:02 examples
-rwxr-xr-x  1 Jason   admin   47500 24 Apr 16:02 libjri.jnilib
-rwxr-xr-x  1 Jason   admin     833 24 Apr 16:02 run
Jason-Bells-MacBook-Pro:jri Jason$
```

You'll set up a basic Eclipse project, and then you can see how the parts fit together. I developed the example on the macOS operating system, but the variations on the other operating systems, such as Windows or Linux, are not that different.

Setting Up an Eclipse Project

Create a Java project and call it JRITest. Go to the properties, click the Java Build Path, and add an external `jar` file. Now look for the `JRI.jar` file, which is normally located in the `/Library/Frameworks/R.framework/Resources/library/rJava/jri` folder. (See Figure 14.11.)

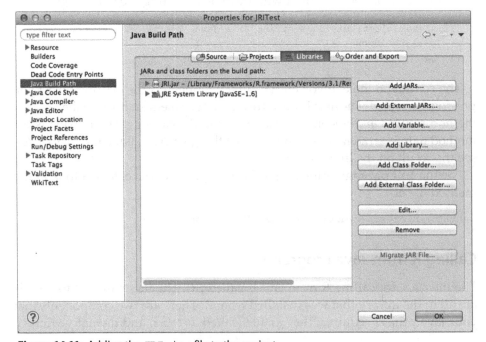

Figure 14.11: Adding the `JRI.jar` file to the project

To make sure the R engine is working within Java, you're going to create a small test file to initialize the engine, load the built-in `iris` dataset, and iterate through an evaluation.

Creating the Java/R Class

To create a new class, select File ⇨ New ⇨ Class and call the new file `TestR.java`.

```
import java.util.Enumeration;

import org.rosuda.JRI.REXP;
import org.rosuda.JRI.RVector;
import org.rosuda.JRI.Rengine;

public class TestR {

    public static void main(String[] args) {
        Rengine rEngine = new Rengine(new String[] { "--vanilla" },
false, null);
        System.out.println("Waiting for R to create the engine.");

        if (!rEngine.waitForR()) {
            System.out.println("Cannot load R engine.");
            return;
        }

        rEngine.eval("data(iris)", false);
        REXP exp = rEngine.eval("iris");
        RVector vector = exp.asVector();
        System.out.println("Outputting data:");
        for (Enumeration e = vector.getNames().elements();
e.hasMoreElements();) {
            System.out.println(e.nextElement());
        }
    }
}
```

The first thing that happens within the `main` method is to start up an R engine. Nothing works until this step is complete.

Next, you pass an R command to load the `iris` data using the `REngine.eval` function. Last, you initialize an R vector within Java, convert the `iris` data to a vector, and then iterate the output.

Running the Example

If you attempt to run the class now, you will get an error from Eclipse, because the R runtime library isn't linked to the project. You get the following error if the library isn't linked:

```
Cannot find JRI native library!
Please make sure that the JRI native library is in a directory listed
```

```
in java.library.path.

java.lang.UnsatisfiedLinkError: no jri in java.library.path
    at java.lang.ClassLoader.loadLibrary(ClassLoader.java:1764)
    at java.lang.Runtime.loadLibrary0(Runtime.java:823)
    at java.lang.System.loadLibrary(System.java:1044)
    at org.rosuda.JRI.Rengine.<clinit>(Rengine.java:19)
    at TestR.main(TestR.java:10)
```

To set that up, you need to look at the run configurations for the project. Select Run ➪ Run Configurations and then click the TestR class. On the Arguments tab, you need to add a -D flag to the virtual machine arguments, as shown in Figure 14.12.

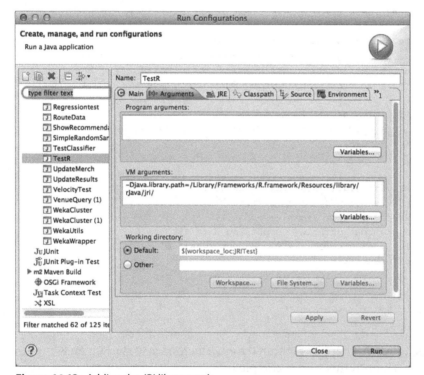

Figure 14.12: Adding the JRI library path

If you get an error about the R _ HOME path not being set, then reopen the run configuration and click the Environments tab, as shown in Figure 14.13. If you use Windows, locate the R.dll file on your system and add it to your path.

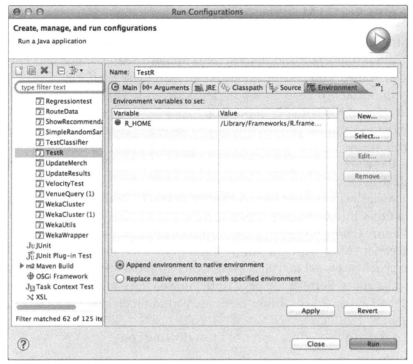

Figure 14.13: Adding the environment R_HOME path

Click Run and try again. This time you should see the correct output.

```
Waiting for R to create the engine.
Outputting names:
Sepal.Length
Sepal.Width
Petal.Length
Petal.Width
Species
```

Extending Your R Implementations

The code you've just walked through gives you the basic framework for getting R functions and commands working from a Java program. The R examples you've seen in this chapter could be easily converted to a Java program using this method if you want. If you have an R expert on your team, then it might be prudent to have that person write the R functions first and then port them to Java.

Connecting to Social Media with R

I know, another mention of the Twitter application program interface (API). This sort of data is important and not just for sentiment analysis. The majority of news, social graphs, and conversations happen on these platforms, and it's increasingly important to keep on top of developments. This goes for R, too; being able to connect to Twitter is important.

In the previous examples you used Twitter's development account to create the authorization keys for the application. You're essentially forcing your application to use these pre-existing keys. That works fine, but you have to approach things a little differently for the twitteR library.

Make a note of the consumer key and secret as you'll need those, but there are couple of other things you need to confirm.

First, make sure there's no callback URL. You confirm this on the Settings tab of your application. Make sure the Allow This Application to Be Used to Sign In to Twitter check box is set to true as well. Don't forget to update the settings for them to take effect. Refresh the page until you see the settings are correct.

Now that the Twitter side of things is set up, you can look at some R code to help you log in to Twitter.

You need to open a text editor and use the following code:

```
library(twitteR)
cred <- OAuthFactory$new(consumerKey="xxxxxxxxxxxxx",
  consumerSecret="xxxxxxxxxxxx",
  requestURL="http://api.twitter.com/oauth/request_token",
  accessURL="http://api.twitter.com/oauth/access_token",
  authURL="http://api.twitter.com/oauth/authorize")
download.file(url="http://curl.haxx.se/ca/cacert.pem",
destfile="cacert.pem")
cred$handshake(cainfo="cacert.pem")
```

Replace the consumer key and secret values with the ones you have for your application. Save the file as `twitterconnect.r` and then quit the text editor.

Back in the R command line, you load the source file, and it runs as soon as it's loaded.

```
>source(twitterconnect.r')
```

You see the twitteR library load the dependencies and then attempt to connect to Twitter.

```
> source("twitterconnect.r")
Loading required package: ROAuth
Loading required package: RCurl
Loading required package: bitops
Loading required package: digest
```

```
Loading required package: rjson
trying URL 'http://curl.haxx.se/ca/cacert.pem'
Content type 'text/plain' length 251338 bytes (245 Kb)
opened URL
==================================================
downloaded 245 Kb
To enable the connection, please direct your web browser to:

http://api.twitter.com/oauth/authorize?oauth_token=cv88UGfPrAJnraJPqdGje
g5QJEUMk185jOUncJhDk

When complete, record the PIN given to you and provide it here:
```

The main thing to look out for is the connecting URL with the `oauth_token` key. Copy the whole URL and paste it into a browser. You go to the Twitter site where you're asked to authorize your application request. When you accept this, you'd normally return to the application, but because you've disabled that feature, you get a personal identification number (PIN) instead.

Back at the R command line, the program you have written is currently waiting for input—the PIN—so type that in the R window.

As soon as the PIN is entered, you're returned to the R prompt. You need to register the oauth with the twitteR library, which you do with the following command:

```
>registerTwitterOAuth(cred)
```

The R command line responds with TRUE when the credentials are registered. After that's done, you can run a quick test.

```
searchTwitter("#bigdata")
```

You start to see results come through to the command line.

```
[[1]]
[1] "eriksmits: #BigData could generate millions of new jobs
http://t.co/w1FGdxjBI9 via @FortuneMagazine, is a Java Hadoop developer
key for creating value?"
[[2]]
[1] "MobileBIAus: RT @BI_Television: RT @DavidAFrankel: Big Data
Collides with Market Research http://t.co/M36yXMWJbg #bigdata
#analytics"
[[3]]
[1] "alibaba_aus: @PracticalEcomm discusses how the use of #BigData can
combat #ecommerce fraud http://t.co/fmd9m0wWeH"
[[4]]
[1] "alankayvr: RT @ventanaresearch: It's not too late! Join us in S.F.
for the 2013 Technology Leadership Summit - sessions on #BigData #Cloud
&amp; more http. . ."
[[5]]
```

```
[1] "DelrayMom: MT @Loyalty360: White Paper 6 #Tips for turning
#BigData into key #insights, http://t.co/yuNRWxzXfZ, @SAS, #mktg #data
#contentmarketing"
```

It's worth saving the credentials so you don't have to keep re-authorizing the access tokens via Twitter. You can save them to a file.

```
>save(cred,file="credentials.RData")
```

The next time you want to use the Twitter credentials, you can do it in two lines.

```
>load("credentials.RData")
>registerTwitterOAuth(cred)
```

Summary

R is a complex piece of software, but it's well-loved by data scientists (new and old ones alike), and also it's become the de facto statistics software for the open source generation.

There are hundreds of well-developed and documented libraries for R within the CRAN package library.

Development does come at a cost; it can be memory intensive for large and complex jobs. It doesn't handle huge amounts of data well, but this is improved by the work done by Revolution Analytics and Purdue University bringing Hadoop processing power to the R engine.

If you still prefer the comfort of your "normal" programming language, such as Java, then use the JRI/RJava combinations; they work very well. Think about real-time analytics to your Java web servlets, for example—very powerful indeed.

Regardless of whether you're processing social media feeds, evaluating e-commerce shopping baskets, or reading sensor data from temperature gauges, look at R as an alternative for processing the data.

Kafka Quick Start

For a full explanation of how Kafka works and the full walk-through of producers and consumers, please look at Chapter 12.

You can download the open source version of Kafka from this link:

```
https://www.apache.org/dyn/closer.cgi?path=/kafka/2.2.0/
kafka_2.12-2.2.0.tgz
```

Installing Kafka

The downloaded distribution comes with both Zookeeper and Kafka and all the command-line tools explained in this appendix. Before you can use it, create a directory to work from and untar the distribution file.

```
$ tar -zxf kafka_2.12-2.2.0.tgz
$ cd kafka_2.12-2.2.0
```

Starting Zookeeper

For development, a stand-alone cluster will work fine, so there are no changes required to any of the configuration files. The first thing to start is Zookeeper.

In a terminal window, run the following as the root user (if you don't use the root user, you may get file permission issues):

```
$ bin/zookeeper-server-start.sh config/zookeeper.properties
[2019-06-18 07:43:12,238] INFO Reading configuration from:
config/zookeeper.properties (org.apache.zookeeper.server.quorum.
QuorumPeerConfig)
[2019-06-18 07:43:12,241] INFO autopurge.snapRetainCount set to 3
(org.apache.zookeeper.server.DatadirCleanupManager)
[2019-06-18 07:43:12,241] INFO autopurge.purgeInterval set to 0
(org.apache.zookeeper.server.DatadirCleanupManager)
[2019-06-18 07:43:12,243] INFO Purge task is not scheduled.
(org.apache.zookeeper.server.DatadirCleanupManager)
[2019-06-18 07:43:12,243] WARN Either no config or no quorum defined in
config, running  in standalone mode
(org.apache.zookeeper.server.quorum.QuorumPeerMain)
[2019-06-18 07:43:12,260] INFO Reading configuration from:
config/zookeeper.properties (org.apache.zookeeper.server.quorum.
QuorumPeerConfig)
```

Starting Kafka

Once Zookeeper is running, open a new terminal window and start the Kafka broker with the following command:

```
$ bin/kafka-server-start.sh config/server.properties
[2019-06-18 07:43:45,792] INFO Registered kafka:type=kafka.Log4jController
MBean
(kafka.utils.Log4jControllerRegistration$)
[2019-06-18 07:43:46,536] INFO starting (kafka.server.KafkaServer)
[2019-06-18 07:43:46,538] INFO Connecting to zookeeper on localhost:2181
(kafka.server.KafkaServer)
[2019-06-18 07:43:46,562] INFO [ZooKeeperClient] Initializing a new
session to localhost:2181. (kafka.zookeeper.ZooKeeperClient)
[2019-06-18 07:43:46,568] INFO Client environment:zookeeper.version=3.4.13-
2d71af4dbe22557fda74f9a9b4309b15a7487f03,
built on 06/29/2018 00:39 GMT (org.apache.zookeeper.ZooKeeper)
[2019-06-18 07:43:46,568] INFO Client environment:
host.name=192.168.1.102 (org.apache.zookeeper.ZooKeeper)
[2019-06-18 07:43:46,568] INFO Client environment:
java.version=1.8.0_45 (org.apache.zookeeper.ZooKeeper)
```

Creating Topics

To write messages to Kafka, you require a topic. To create a topic from the command line, use the following command. Note that the minimum replication factor and partition count is 1. You can increase it if required in the future, but you cannot reduce those figures once set; the only way to do that is to delete the topic and start again.

```
$KAFKA_HOME/bin/kafka-topics.sh --zookeeper localhost:2181 --create
--topic testtopic --replication-factor 1 --partitions 1
Created topic "testtopic".
```

Listing Topics

To list the topics, use the `--list` flag from the `kafka-topics` command. Note that all topics will be listed including the internal ones used by Kafka for offset counts and metrics.

```
bin/kafka-topics --zookeeper localhost:2181 --list
__confluent.support.metrics
__consumer_offsets
_confluent-command
_confluent-metrics
_confluent-monitoring
_schemas
connect-configs
connect-offsets
connect-statuses
testtopic
```

Describing a Topic

The describe topic flag, `--describe`, will output to the console the general information plus any other added configuration settings of a given topic.

```
$ bin/kafka-topics --zookeeper localhost:2181 --describe
--topic testtopic
Topic:testtopic    PartitionCount:1    ReplicationFactor:1    Configs:
    Topic: testtopic    Partition: 0    Leader: 0    Replicas: 0
Isr: 0
```

Deleting Topics

Deleting a topic is done within the same `kafka-topics` command with the `--delete` flag. If your broker or topic configuration contains `delete.topic.enable=false`, then your request will be ignored, but no warning will be given. Depending on the size of your topic log, it may not delete immediately; it's worth checking again with the `--list` flag to confirm.

```
$ bin/kafka-topics --zookeeper localhost:2181 --delete --topic testtopic
Topic testtopic is marked for deletion.
Note: This will have no impact if delete.topic.enable is not set to
true.
```

Running a Console Producer

You are able to create test messages from the command line using the `kafka-console-producer` script. When it is run, you will be able to either type or copy/paste information into the terminal window, and it will be sent to the specified topic.

```
$ bin/kafka-console-producer --broker-list localhost:9092
--topic testtopic
```

Running a Console Consumer

To read the messages from a topic from the terminal, use the `kafka-console-consumer` command.

```
$ bin/kafka-console-consumer --bootstrap-server localhost:9092
--topic testtopic --from-beginning
```

The Twitter API Developer

Application Configuration

Before you can start consuming data from Twitter into an application, you need to set up a development account and install the Twitter development environment. This setup process uses Twitter's developer site (http://dev.twitter.com) and requires you to have an existing Twitter account. If you don't have a Twitter account, you can sign up for one at http://www.twitter.com.

The developer website enables users to create keys for their applications to access Twitter's APIs. I'm going step-by-step, assuming you've never done it before. If you already know how to create Twitter developer access and the required credentials, then you can skip the remainder of this section.

1. Open a web browser and go to http://dev.twitter.com. Log in with your Twitter credentials.

2. Find your Twitter avatar image at the top right. Hover your mouse pointer over the arrow next to it. Click the My Applications link in the drop-down menu.

3. Click the Create A New Application button, as shown in Figure B.1.

4. Fill in the required fields for the application name, the description, and a URL for your website (see Figure B.2). This information is required if Twitter users decide to use their accounts with your application. You can leave the callback URL blank, because you won't be using it.

You need to agree to Twitter's terms and conditions as well as fill in a captcha code. Click the Create Your Twitter Application button.

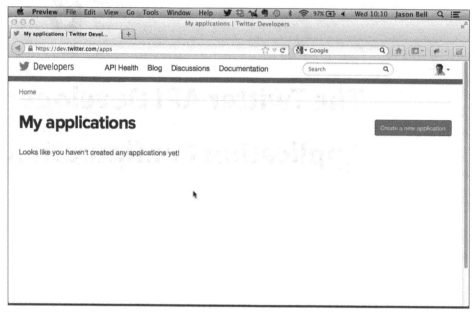

Figure B.1: Creating a new Twitter application page

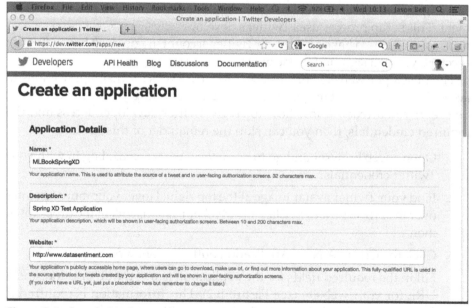

Figure B.2: Completing the application detail page

Assuming that all went well, you are directed to the Details configuration screen of your application. These settings include organization, authorization, settings, access keys, and an option to delete the application.

5. In the OAuth Settings area of the Details tab (see Figure B.3) is the information you require for the Spring XD system to collect Twitter streaming data. Make a note of the consumer key and the consumer secret.

6. Twitter requires an access token to go with your consumer key and secret. The easiest way to get one is to click the Create My Access Token button below the OAuth settings. It can take a moment for the access token to appear in the Details page. If it takes longer than 30 seconds or if the access token remains empty, refresh the page in your browser, and the tokens appear.

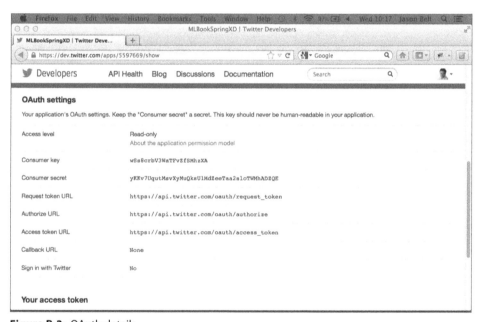

Figure B.3: OAuth details

When the process is complete, you will see the access token, an access token secret key, and the access level. A read-only token is fine for your purposes, because you'll be consuming data and not writing new tweets.

Useful Unix Commands

Regardless of the operating system you use on a daily basis, there's nothing wrong with learning some handy Unix commands. The commands covered in this section will help you on the day-to-day tasks of quickly testing, parsing, and searching through your text data.

If you're a Windows user, you can still join in by downloading Cygwin, which is a Unix shell command interpreter. Cygwin is a shell that sits on top of the Windows Command application and behaves like it's a Unix install.

Some of these commands will appear from time to time in the chapters of this book, so it's worth reviewing them now. Experiment with them and study the output so you have an idea of what to expect.

Using Sample Data

Before you get started with the tools, you need some sample data. With a text editor, type out the following lines and separate each value with a tab:

```
987    1391548780    hhh bbb
988    1391548781    sda jjj
989    1391548782    asd asd
990    1391548783    gjh jkl
991    1391548784    abc abc
992    1391548785    ghj gjh
```

```
993     1391548785    hhh bbb
994     1391548785    sda jjj
995     1391548786    asd asd
996     1391548787    gjh jkl
997     1391548787    abc abc
998     1391548787    ghj gjh
```

Name the text file `text.txt`, and then you can follow along with the following commands and see the output. The sample data is basically comprised of a unique ID, a timestamp, and some text. It's the sort of thing you would see within a database table but is output as text. This example uses a tab delimiter, but you might find data with commas, semicolons, and other characters used as a delimiter.

Showing the Contents: cat, more, and less

The `cat` command concatenates and prints the contents of one or more files to the console output.

Example Command

```
cat text.txt text2.txt text3.txt
```

Expected Output

```
987     1391548780    hhh bbb
988     1391548781    sda jjj
989     1391548782    asd asd
990     1391548783    gjh jkl
991     1391548784    abc abc
992     1391548785    ghj gjh
993     1391548785    hhh bbb
994     1391548785    sda jjj
995     1391548786    asd asd
996     1391548787    gjh jkl
997     1391548787    abc abc
998     1391548787    ghj gjh
```

Adding `-b` to the options gives you the line numbers, too.

```
$ cat -b sample.txt
     1  987     1391548780    hhh bbb
     2  988     1391548781    sda jjj
     3  989     1391548782    asd asd
```

```
 4 990    1391548783    gjh jkl
 5 991    1391548784    abc abc
 6 992    1391548785    ghj gjh
 7 993    1391548785    hhh bbb
 8 994    1391548785    sda jjj
 9 995    1391548786    asd asd
10 996    1391548787    gjh jkl
11 997    1391548787    abc abc
12 998    1391548787    ghj gjh
```

If too much content is showing in the console for you to keep up with, you can add the `more` command after the `cat` command.

```
$ cat sample | more
```

Alternatively, the `less` command gives you a controlled environment for viewing the contents of files but does not let you edit them.

Filtering Content: grep

For matching patterns within text, `grep` is your friend. It's one of those utilities you'll use again and again after you get used to it. The syntax is very basic: `grep [options] [pattern to find] [name of file(s)]`.

Example Command for Finding Text

```
$grep 'bbb bbb' sample.txt
```

Example Output

To invert the output, find the lines that don't match the pattern and then add the `-v` flag before the pattern. Table C.1 shows other handy option flags you can use with `grep`.

Table C.1: grep Option Flags

FLAG	EFFECT ON OUTPUT
-c	Outputs the number of times the pattern was matched in the file
-v	Inverts the output to show lines that don't match the pattern
-i	Ignores the case on the input line so BBB and bbb would match
-n	Displays the line number on which the pattern match occurs
-l	Lists the filenames where the pattern matches

The pattern matching can be taken a step further by introducing Perl-like patterns with the -P option. There are numerous books on the subject of Perl and regular expressions.

```
$ grep -e '[1-5]\tsda' sample.txt
988     139154878    sda jjj
994     1391548785   sda jjj
```

Sorting Data: sort

The sort command takes the input file and sorts in ascending or descending order. By default, sort assumes that everything is a string. (You'll read about number values in a moment.)

Example Command for Basic Sorting

```
$sort sample.txt
```

Example Output

```
987     1391548780   hhh bbb
988     1391548781   sda jjj
989     1391548782   asd asd
990     1391548783   gjh jkl
991     1391548784   abc abc
992     1391548785   ghj gjh
993     1391548785   hhh bbb
994     1391548785   sda jjj
995     1391548786   asd asd
996     1391548787   gjh jkl
997     1391548787   abc abc
998     1391548787   ghj gjh
```

The -r flag outputs the results in descending order.

```
$ sort -r sample.txt
998     1391548787   ghj gjh
997     1391548787   abc abc
996     1391548787   gjh jkl
995     1391548786   asd asd
994     1391548785   sda jjj
993     1391548785   hhh bbb
992     1391548785   ghj gjh
991     1391548784   abc abc
990     1391548783   gjh jkl
989     1391548782   asd asd
```

```
988     1391548781    sda jjj
987     1391548780    hhh bbb
```

So far, the example has covered sorting strings. Consider the following list of numbers:

```
10
18
1
20
17
15
103
110
12
22
21
201
```

Running a default `sort` command would sort the number values as strings, so the output would be correct, but it probably would not be what you hoped for.

```
$ sort sample2.txt
1
10
103
110
12
15
17
18
20
201
21
22
```

The `-n` option treats the input data as numeric.

```
$ sort -n sample2.txt
1
10
12
15
17
18
20
21
22
103
110
201
```

The final useful option is the -k flag, which splits the input data into columns and lets you sort on a specific column. In the sample data, the first three-letter text column is the third, so by using the option -k3, you sort on the text column.

```
$ sort -k3 sample.txt
991     1391548784    abc abc
997     1391548787    abc abc
989     1391548782    asd asd
995     1391548786    asd asd
992     1391548785    ghj gjh
998     1391548787    ghj gjh
990     1391548783    gjh jkl
996     1391548787    gjh jkl
987     1391548780    hhh bbb
993     1391548785    hhh bbb
988     1391548781    sda jjj
994     1391548785    sda jjj
```

When combining the sort option flags, you can output pretty much anything you want. For example, you could reverse sort the second column as a numeric value using the following code:

```
$ sort -k2nr sample.txt
996     1391548787    gjh jkl
997     1391548787    abc abc
998     1391548787    ghj gjh
995     1391548786    asd asd
992     1391548785    ghj gjh
993     1391548785    hhh bbb
994     1391548785    sda jjj
991     1391548784    abc abc
990     1391548783    gjh jkl
989     1391548782    asd asd
988     1391548781    sda jjj
987     1391548780    hhh bbb
```

The combination of the -k, -n, and -r flags is suitable for most applications when you're running through basic delimited data.

Finding Unique Occurrences: uniq

Sometimes data will have repeat lines in them because of a data entry error or a slightly error-prone database query. Before raising your voice at the database operator (I have to look in the mirror to talk to my DBA), you can use the uniq command to extract all the unique lines.

```
$ uniq sample.txt
987     1391548780    hhh bbb
```

```
988    1391548781    sda jjj
989    1391548782    asd asd
990    1391548783    gjh jkl
991    1391548784    abc abc
992    1391548785    ghj gjh
993    1391548785    hhh bbb
994    1391548785    sda jjj
995    1391548786    asd asd
996    1391548787    gjh jkl
997    1391548787    abc abc
998    1391548787    ghj gjh
```

Using uniq in combination with the -c flag, the output will contain the number of repeats that were found within the file. If you do not need to see the counts, then you can use the -u flag in the sort command.

Showing the Top of a File: head

When you've downloaded a 10,000-line file and you want to quickly inspect it, it's easy to use the cat command and dump the whole file out. Obviously, you can't read it as the lines are appearing faster than your eyes. The head command shows the output of a defined number of lines.

```
$ head -n 5 sample.txt
987    1391548780    hhh bbb
988    1391548781    sda jjj
989    1391548782    asd asd
990    1391548783    gjh jkl
991    1391548784    abc abc
```

If you want to show the last number of lines of the file, then use the tail command.

Counting Words: wc

Using wc counts the number of lines, words, and characters used within the text document.

```
$ wc sample.txt
    13      48     277 sample.txt
```

If you want to see only the number of lines, then use -l as an option, -w for words, and -m for characters.

Locating Anything: find

The find command is one of my favorites. It takes a little getting used to, but once you do, it will be a big help to you.

The syntax starts with a source directory (using a period [.] if you want to start in the current directory) and various text options. For example, here's what it looks like to run the find command on the home directory on my machine:

```
$ find . -type f -name '*.txt' -print
././gradle/caches/1.10/scripts/build_361heej71i7errcd096649rjns/
ProjectScript/buildscript/classes/emptyScript.txt
././gradle/caches/1.8/scripts/build_4p3rjceekul5gbo8of3d9h4hso/
ProjectScript/buildscript/classes/emptyScript.txt
././rvm/config/displayed-notes.txt
././rvm/gems/ruby-1.9.2-p320/gems/arel-3.0.2/History.txt
././rvm/gems/ruby-1.9.2-p320/gems/arel-3.0.2/Manifest.txt
```

With the -type flag, I am looking for files (f). -name gives the types of file I'm looking for—in this instance any file with the extension .txt. I print the results with the -print flag.

So far, there's nothing here that a few Unix commands couldn't do. The great thing about find is that you can use Unix commands within the command itself. Imagine you're looking for any files with the phrase "SOFTWARE IS PROVIDED" in the body of the text file.

```
$ find . -type f -name "*.txt" -exec grep "SOFTWARE IS PROVIDED" {} \;
-print
THE SOFTWARE IS PROVIDED "AS IS", WITHOUT WARRANTY OF ANY KIND,
././rvm/gems/ruby-1.9.2-p320/gems/arel-3.0.2/MIT-LICENSE.txt
THE SOFTWARE IS PROVIDED "AS IS", WITHOUT WARRANTY OF ANY KIND,
././rvm/gems/ruby-1.9.2-p320/gems/polyglot-0.3.3/License.txt
THE SOFTWARE IS PROVIDED "AS IS", WITHOUT WARRANTY OF ANY KIND,
././rvm/gems/ruby-1.9.2-p320/gems/polyglot-0.3.3/README.txt
THE SOFTWARE IS PROVIDED "AS IS", WITHOUT WARRANTY OF ANY KIND,
EXPRESS OR
././rvm/gems/ruby-1.9.2-p320/gems/rack-test-0.6.1/MIT-LICENSE.txt
THE SOFTWARE IS PROVIDED "AS IS", WITHOUT WARRANTY OF ANY KIND,
EXPRESS OR
././rvm/gems/ruby-1.9.2-p320/gems/rack-test-0.6.2/MIT-LICENSE.txt
THE SOFTWARE IS PROVIDED "AS IS", WITHOUT WARRANTY OF ANY KIND,
././rvm/gems/ruby-1.9.2-p320/gems/uglifier-1.2.6/LICENSE.txt
THE SOFTWARE IS PROVIDED "AS IS", WITHOUT WARRANTY OF ANY KIND,
././rvm/gems/ruby-1.9.2-p320/gems/uglifier-1.3.0/LICENSE.txt
```

This example uses the standard grep command that is covered earlier in this appendix, but it replaces the filename with {}, which is a placeholder for the filename on which the find command is working.

find has saved my programming sanity many times over, and if I'm looking for a file that has a method name but I can't remember where it is (this happens a lot with large codebases), I don't have to fire up my IDE. I just use find from the command line.

Combining Commands and Redirecting Output

The explanations for the Unix commands showed you how to use one command at a time. Using the pipe symbol (|), you can chain these commands together. With the greater-than symbol (>), you can redirect output to a new file.

```
$ grep 'hhh' sample.txt | sort | less
987    1391548780    hhh bbb
993    1391548785    hhh bbb
```

The preceding command uses grep to search for 'hhh' in the sample.txt file and then runs the result through the sort command; this is then written to a new file called newsample.txt. Finally, the output of the new file is sent to the console with cat.

Picking a Text Editor

The choice of text editor is a personal one, akin to liking a specific music group, sports team, or favorite actor or actress. I'm not saying fights have broken out about these editors, but conversations over coffee and beer have sometimes been emotional. If you want to work out the personality type of a software developer, just ask him what text editor he uses.

Ultimately, it's up to you what sort of text editor you'll use. You might not be using it all the hours of the day, but for the times you need to quickly look at or edit something, you'll want an editor that you can use fluently and without fuss.

Colon Frenzy: Vi and Vim

Vi was written in 1976 by Bill Joy and was introduced in 1978 when it was released as part of BSD Unix. Pronounced "vee eye," it's been the mainstay of Unix system administrators for a long time. It's still one of the most widely used text editors today.

The thing that makes vi challenging is the concept of *modes* for the arcane commands: There are line-oriented "ex" commands that operate on lines of text; they are needed to write the file (:w) or quit the application (:q). If you want to quit without saving, then :q! is needed. Likewise, overwriting an existing file uses :w!. There is an insert mode while you are entering new text, and there is an edit mode for moving and changing text. Vi is so simple there is a coffee mug available with all (100 percent) of the vi commands on it.

It takes time to learn the commands, such as how to delete lines, yank text into "buffers" and then paste it, search for text, and do other basic functions. After a bit of time, the modes actually become second nature. Vi includes a powerful "macro language" for adding your own commands or command sequences.

I'm still a user of vi; it covers most of my needs. However, I have to admit it drove me mad when I first used it in 1995. I've still not moved over to vim (Vi iMproved), though it has windows, better editing, mouse support, and colorful syntax highlighting.

Nano

For those who can't cope with the colon command, the nano editor provides command-line heaven without all the complication. The nano editor lets you navigate around using the arrow keys (vi uses the h, j, k, and l keys for navigating). The editor includes a decent number of hints at the bottom of the screen, so you can see how you can save and quit.

For sheer speed of loading, editing, saving, and exiting a document, nano is a good choice.

Emacs

Compared to vi and nano, Emacs is the big gun of the text editors. It's highly customizable and configurable. With the resurgence in Lisp-like languages, Clojure being an example, the usage of Emacs is on the rise. For all my Clojure development, I use Emacs. Along with CIDER, it gives me editing, a REPL, and debugging all in one place.

If you want to try Emacs, I recommend the xemacs version of the program.

Emacs bypasses the whole insert mode issue that vi users are willing to adopt. Like nano, in Emacs you type on the screen and see the changes. It also has more than 2,000 built-in commands for those who are happy to use Emacs as their main editor for everything. It is extensible by writing Lisp programs. To exit Emacs if you start it accidentally, press Ctrl-X and then Ctrl-C.

Further Reading

Machine learning is only part of the story; it's the application of knowing what to use to get the insight you need. The domain of data science combines several disciplines that cover programming, math, domain knowledge, and visualization.

It's rare for one book to cover it all. To that end, I've included some further reading that will be of help to you on your machine learning and data journey. (I know what you're thinking, and yes, I have bought and read all of these books.)

Machine Learning

The machine learning arena is a huge domain and the majority of the books written are big, in-depth, heavy affairs that can take time to read, digest, and appreciate. Two stand out:

Data Mining – Practical Machine Learning Tools and Techniques by Ian H. Witten, Eibe Frank, and Mark A. Hall (Morgan Kaufmann, 2011, ISBN 9780123748560)

Collective Intelligence in Action by Satnam Alag (Manning, 2008, ISBN 9781933988313)

Statistics

More and more emphasis is being put on statistical knowledge and its application. Sometimes it feels hard to get into, especially for software developers, so these titles will help you along:

The Art of Statistics by David Spielgalhalter (Pelican Books, 2019, ISBN 9780241398630)

Naked Statistics: Stripping the Dread from the Data by Charles Wheelan (Norton, 2013, ISBN 9780393071955)

Keeping Up with the Quants: Your Guide to Understanding and Using Analytics by Thomas H. Davenport and Jinho Kim (Harvard Business Review Press, 2013, ISBN 9781422187258)

Big Data and Data Science

Regardless of whether you are a supporter of the term Big Data, there's no denying the impact that data has on industry. In Big Data, planning is key, and it's important to have a proper understanding of the implications of planning and insight.

Data Just Right: Introduction to Large-Scale Data & Analytics by Michael Manoochehri (Addison-Wesley, 2014, ISBN 9780321898654)

Big Data: Understanding How Data Powers Big Business by Bill Schmarzo (Wiley, 2013, ISBN 9781118739570)

Big Data @ Work: Dispelling the Myths, Uncovering the Opportunities by Thomas H. Davenport (Harvard Business Review Press, 2014, 9781422168165)

Big Data: A Revolution That Will Transform How We Live, Work, and Think by Viktor Mayer-Schonberger and Kenneth Cukier (Eamon Dolan/Houghton Mifflin Harcourt, 2013, ISBN 9780544002692)

Data Smart: Using Data Science to Transform Information into Insight by John W. Foreman (Wiley, 2013, ISBN 9781118661468)

Data Science for Business: What You Need To Know About Data Mining and Data-Analytic Thinking by Foster Provost and Tom Fawcett (O'Reilly Media, 2013, ISBN 9781449361327)

Visualization

My book concentrates on the pure back-end processing of data with machine learning techniques, but do not discount the power of visualization to communicate your results. These books will help:

Visualize This: The Flowing Data Guide to Design, Visualization, and Statistics by Nathan Yau (Wiley, 2011, ISBN 9780470944882)

Information Is Beautiful by David McCandless (Harper Collins, 2012, ISBN 9780007492893)

Facts Are Sacred by Simon Rogers (Faber & Faber, 2013, ISBN 9780571301614)

Making Decisions

The key to machine learning projects is making good decisions. With insight in hand, you can form next steps. The books listed here aren't software oriented at all, but they will give you vast pools of thinking about how to process and make decisions with the information you have.

Thinking in Bets by Annie Duke (Portfolio/Penguin, 2018, ISBN 9780735216358)

How to Develop Your Thinking Ability by Kenneth S. Keyes, Jr. (McGraw/Hill Paperbacks, 1963)

> A brilliant book, now out of print but can be found by various sellers on the Internet. If it's good enough for Marilyn Monroe, it's good enough for me.

Eyes Wide Open by Noreena Hertz (HarperCollins, 2013, ISBN 9780062268617)

The Signal and the Noise: Why So Many Predictions Fail—but Some Don't by Nate Silver (Penguin Books, 2012, ISBN 9781594204111)

Lean Analytics: Use Data to Build a Better Startup Faster by Alistair Croll and Benjamin Yoskovitz (O'Reilly Media, 2013, ISBN 9781449335670)

Obviously Awesome by April Dunford (Ambient Press, 2019, ISBN 9781999023003)

> Not strictly a decision-making book, but if you are a startup doing anything, then this is the book you need.

Datasets

Sometimes it's hard to find data to play with. Luckily, there are a few websites with loads of the stuff to download.

- **UCI Machine Learning Repository:** http://archive.ics.uci.edu/ml/

  The UCI maintains 290 datasets covering many different domains. What's the most popular downloaded dataset? It's still the iris.

- **Quora:** http://www.quora.com/Where-can-I-find-large-datasets-open-to-the-public

  Here you'll find a long list of URLs covering all sorts of topics that you can investigate. (This site requires you to sign in.)

- **Kaggle:** https://www.kaggle.com

  There are plenty of datasets, advice, tutorials, and competitions to pour through here.

Blogs

And they said RSS feeds were dead. . .I don't think so! There are a few blogs that I keep an eye on regularly, and these are the ones that relate to what is covered in this book:

- **FiveThirtyEight:** http://www.fivethirtyeight.com

  Nate Silver and a team of contributors build this daily digest of stories with data, covering everything from politics to which is the best burrito in the United States.

- **Radar:** http://radar.oreilly.com

  This site for emerging technologies is worth checking out for the daily "Four Short Links," which pinpoints some interesting programs, stories, and case studies from around the Internet.

- **MathBabe:** http://mathbabe.org

  Cathy O'Neill's blog discusses data, quantitative issues, and other subjects within the analytics arena.

Useful Websites

Although Google does a good job of showing you where to find the best sites, I still refer to the following sites when I'm looking for specifics:

- **Wiley:** www.wiley.com

  This is the main website for all Wiley books and also the place to go for the sample code examples for this book.
- **Stack Overflow:** www.stackoverflow.com

  A community of developers helping a community of developers, what's not to like? This site is definitely worth a quick look for answers on coding, servers, and machine learning.

The Tools of the Trade

Here are the links to the tools that are used in this book. It's worth having them bookmarked for updates and announcements.

- **Apache Spark:** http://spark.apache.org
- **Weka:** htwww.cs.waikato.ac.nz/ml/weka
- **Kafka:** http://kafka.apache.org
- **DeepLearning4J:** https://deeplearning4j.org

Index